The Dictionary of

ENVIRONMENTAL MICROBIOLOGY

The Dictionary of

ENVIRONMENTAL MICROBIOLOGY

Linda D. Stetzenbach

Harry Reid Center for Environmental Studies
University of Nevada, Las Vegas
Las Vegas, Nevada

and

Marylynn V. Yates

Department of Environmental Sciences
University of California, Riverside
Riverside, California

ACADEMIC PRESS

An imprint of Elsevier Science

Amsterdam Boston London New York Oxford Paris
San Diego San Francisco Singapore Sydney Tokyo

Cover images: (Clockwise from top left on front cover) Colonies of heterotrophic bacteria on tryptic soy agar, courtesy of P. Jacoby-Garrett, University of Nevada, Las Vegas; Rotavirus viewed with negative stain transmission electron microscopy, courtesy of F. P. Williams, United States Environmental Protection Agency; Immunofluorescently stained *Giardia* cysts viewed epifluorescent microscropy, courtesy of H. D. A. Lindquist, United States Environmental Protection Agency; *Exserohilum* spores and hyphae viewed with light microscopy, courtesy of Patricia Cruz, University of Nevada, Las Vegas. (Back cover) Acridine orange-stained *Pseudomonas* cells viewed with epifluorescent microscopy, courtesy of Linda D. Stetzenbach, University of Nevada, Las Vegas.

This book is printed on acid-free paper. ∞

Academic Press
An imprint of Elsevier Science
525 B Street, Suite 1900, San Diego, California 92101-4495, USA
http://www.academicpress.com

Academic Press
84 Theobald's Road, London WC1X 8RR, UK
http://www.academicpress.com

Library of Congress Catalog Card Number: 2003102215

International Standard Book Number: 0-12-668000-0

PRINTED IN THE UNITED STATES OF AMERICA
03 04 05 06 07 8 7 6 5 4 3 2 1

Preface

"Everything is everywhere, the environment selects"
M. W. Beijerinck (1851–1931)

This statement sets the tone for scientific investigation of the microbial world and stands as a reminder of the ubiquitous presence of microorganisms in the environment. The challenge for environmental microbiologists is in understanding the functional role that viruses, bacteria, fungi, and parasites have in environments as diverse as the depths of the sea and the far reaches of space. This understanding requires knowledge of the basic science of microbiology, including the molecular nature, structure, and function of the organisms, and the applied science of microbiology, including development and use of appropriate sampling, processing, and assay methods for the collection of data, statistical analyses, interpretation of data, and verbal or written reporting.

Entries in this dictionary provide information to assist scientists in the study of environmental microbiology, but this book also serves as a resource for non-scientists. Scientific names of organisms are italicized and terms that are cross-referenced are underlined. Acronyms are cross-referenced to the complete entry for the term.

We acknowledge Paula Jacoby-Garrett for producing the illustrations and Sean Abbott, Roger Fujioka, and Michael Gealt for their critical review of the manuscript and helpful comments. We thank our husbands (Klaus J. Stetzenbach and Scott R. Yates) and our children (Erika A. Stetzenbach Theising, Kristen A. Stetzenbach, Matthew D. Yates, and Teresa C. Yates) for their support throughout the research, writing, rewriting, and editing of this text. When we met as graduate students 20 years ago we never envisioned how our families, our friendship, and our professional careers would intertwine, enriching us all.

Linda D. Stetzenbach
Marylynn V. Yates

α Greek symbol for alpha used to denote the probability that a Type I error will be made; designated as the *P* value in statistical analysis.

ABC active biological containment.

abiogenesis the theory of spontaneous generation of living cells from nonliving material.

abiotic nonbiological.

ABPA allergic bronchopulmonary aspergillosis.

ABS alkyl benzyl sulfonates.

abscissa x-axis.

Absidia fungal genus; rapidly growing colony on malt extract agar that is wooly in texture with a gray surface and uncolored reverse; aseptate hyphae are large, some with adventitious septa; exhibit a few rhizoides that are situated on stolons located between sporangiophores which are branched with an apophysis beneath pyriform sporangia; distinguished from *Rhizopus*, *Rhizomucor*, and *Mucor* by the well-developed apophysis; isolated from soil and decaying vegetation with some pathogenic species, generally involving an immunocompromised patient.

absolute risk the incidence of a disease in a population.

absorbed dose the amount of material that penetrates across a cellular boundary following exposure.

absorption the penetration of material into the interior matrix of another substance; in contrast to adsorption.

abstriction the formation of spores by the growth of septa thereby cutting off successive sections of a conidiogenous cell.

Acanthamoeba protozoan; free-living amoebae commonly found in natural waters.

accession number sequential numbering of specimens according to the order in which they are obtained; also used for nucleic acid sequences in a database such as GenBank.

accessory pigment a pigment (e.g., chlorophyll) involved with photosynthesis that collects light at different wavelengths and transfers the energy to the primary system.

accidental fecal release (AFR) inadvertent discharge of human fecal material into recreational waters such as beaches, swimming pools, and jacuzzi tubs.

accuracy the degree of agreement that a measurement has with its true value; in contrast to precision.

acellular without a cell structure.

acerose stiff and needle-like.

acetic acid bacteria aerobic, gram-negative acid-tolerant organisms that are able to grow at pH <5 and incompletely oxidize organic carbon to alcohol; oxidize ethanol to acetic acid which accumulates in the reaction mixture, in contrast to other aerobic bacteria that oxidize their energy sources completely to CO_2 and water.

Acetobacter bacterial genus; acid tolerant, aerobic, gram-negative bacilli, an acetic acid bacterium that ferments ethanol to acetic acid but also has the needed enzymes to complete the cycle by oxidizing acetic acid to CO_2 and water; commonly used in the commercial production of vinegar.

acetogenesis the formation of acetate from CO_2 as the result of the metabolic processes of certain bacteria, in contrast to methanogenesis; three recognized processes are the acetyl-CoA pathway, the glycine synthase-dependent pathway, and the reductive citric acid cycle.

acetogenic bacteria approximately 40 recognized species isolated from diverse anaerobic habitats that conduct acetogenesis using the acetyl-CoA pathway for the conservation of energy and growth; a variety of bacillus and coccoid-shaped gram-negative and gram-positive bacteria that are mostly mesophilic, although some psychrophilic and thermophilic species have been isolated.

acetone solvent used as a decolorizing agent in the Gram reaction to remove crystal violet from the cell wall of gram-negative bacteria; an extraction solvent for lipids and sterols.

acetotroph/acetotrophic methanogenic bacterium that uses acetate as the carbon source resulting in the conversion to CH_4 and CO_2.

acetyl-CoA pathway autotrophic CO_2 fixation conducted by obligate anaerobes including methanogens and sulfate-reducing bacteria.

acetylene reduction assay method used to assess nitrogen fixation in which N_2 is substituted with acetylene which is reduced to ethylene or ethane.

ACGIH American Conference of Governmental Industrial Hygienists.

achlorophyllous nonphotosynthetic; lacking chlorophyll.

Acholeplasma laidlawii microbial species; a member of the Mollicutes but these organisms do not require sterol or serum; initially isolated from sewage, compost, and soil.

Acholeplasmataceae microbial family; within the order Mollicutes; organisms that lack a cell wall and do not have a growth requirement for cholesterol; in contrast to Mycoplasmataceae.

achromatic lens an objective lens that corrects for chromatic aberration for two colors and spherical aberration for one color; relatively inexpensive lens used for the routine observation of microorganisms and commonly used in introductory microbiology laboratory classes.

acicular needle-shaped.

acid fast / acid alcohol fast characteristic in which bacteria stained with an aniline dye do not decolorize when treated with acid or acid alcohol; property of the genus *Mycobacteria*.

acidic stain reagent with a positive charged, basic chromophore that is attracted to negatively charged cell material.

acid mine drainage sulfuric acid (H_2SO_4)-containing water resulting from the microbial oxidation of iron sulfide minerals.

acidophile microorganism that preferentially grows at an acidic pH; cells that have an affinity for acidic dye.

acid rock drainage (ARD) the bacterial mediated leachate resulting from the oxidation of sulfide minerals exposed to air and water, and the products of the interaction of alkaline rock and water with acidic metal-containing solutions.

acid tolerance response (ATR) assessment of the acid tolerance of bacteria generally conducted as challenge experiments with the addition of lactic or acetic acid to growth media and evaluation of the survival of test organisms.

acidulant an additive used to increase the acidity of a food product.

Acinetobacter bacterial genus; member of the Neisseriaceae; gram-negative, oxidase negative coccobacillus that is saprophytic; isolated in soil, water, sewage, and clinical samples; currently 7 recognized species.

Acinetobacter baumannii bacterial species; isolated from air samples associated with wastewater treatment operations.

Acinetobacter calcoaceticus bacterial species; produces bioemulsifiers that are polysaccharide and protein based; isolated from clinical and environmental samples.

Acinetobacter radioresistens bacterial species; produces a protein-based bioemulsifier.

AcNPV the designation of a nucleopolyhedron baculovirus that infects the alfalfa looper (*Autographa californica*).

Acremonium fungal genus; approximately 80–90 species; colony on malt extract agar is small, membranous or thinly velvety, white, pale pink or salmon in color; small colorless spores form chains or slimy heads; ubiquitous, isolated from soil, organic debris, hay, food stuffs, and very wet indoor conditions; often found growing with *Stachybotrys*; most species do not grow at 37°C; associated with Type I allergies and Type III hypersensitivity; produces the class of antibiotics known as the cephalosporins.

acrid a sharp, bitter, irritating, or stinging odor or taste.

acridine orange (AO) a fluorescent dye used with epifluorescent microscopy direct counting of cells in suspension that intercalates into nucleic acid resulting in apple-green fluorescence of cells and orange fluorescence of cytoplasmic RNA and nucleolar RNA in fungi and other eukaryotes.

acrogenous of or at the tip.

acropetal characteristic of fungal conidia arranged in chains in which the youngest is located at the tip of the chain; in contrast to basipetal.

Actidione cycloheximide.

actinobacteria actinomycetes.

Actinomyces bacterial genus; member of the Actinomycetaceae; colonies are white to gray or creamy white but some species are pigmented pink to red; mature colonies are rough and crumbly or smooth and soft to mucoid; straight or slightly curved, gram-positive, chemoorganotrophic, anaerobic to facultatively anaerobic bacilli with slender filaments that fragment into coryneform cells; may be pathogenic to animals and humans.

Actinomycetaceae bacterial family; taxonomically in the order Actinomycetales; characterized as non-acid fast irregularly shaped cells that do not produce mycelia or spores; representative genera include *Actinomyces, Arachnia, Bifidobacterium, Bacterionema,* and *Rothia.*

Actinomycetales bacterial order; members are characterized by the formation of branching filaments, resembling the fungi, but these organisms are prokaryotic.

actinomycetes bacteria in the family Actinomycetaceae; filamentous, branching bacteria that comprise approximately 10–33 percent of the classified microbial population in soils although the use of rRNA sequencing may result in the lowering of this percentage; characterized by the formation of branching or true filaments; relatively resistant to desiccation; many produce antibiotics; also termed actinobacteria.

Actinoplanaceae bacterial family; taxonomically in the order Actinomycetales; characterized as saprophytes or facultative parasites that form mycelia and bear spores within sporangia; representative genera include *Actinoplanes, Amorphosoporangium, Ampullariella, Dactylosporangium, Kitasatoa, Pilimelia, Planobispora, Planomonospora, Spirillospora,* and *Streptosporangium.*

action level the amount or concentration of a contaminant in food, water, soil, or air that specifies the initiation of an action or countermeasure to minimize the potential for adverse effects on health or environmental quality.

activated sludge an aerobic secondary wastewater treatment process that relies on microbial growth to reduce nutrient and organic content in sewage.

activation the first stage of germination of endospores followed by germination and outgrowth.

activation energy the threshold energy level required to initiate a chemical or biological reaction.

active biological containment (ABC) control of the survival of bacterial populations with a sensory system that recognizes physical or chemical signals in the environment.

active fungal spores dispersal release of fungal spores as a result of a discharge mechanism present in some fungi; activation of the dispersal is generally associated with increased humidity or moisture events; in contrast to passive fungal spore dispersal.

active sampling collection of material using a mechanical device or instrument.

acquired immune deficiency syndrome (AIDS) an infectious disease resulting from infection with the human immunodeficiency virus (HIV).

acuminate gradually tapering at the tip.

acute having a sudden onset; may be the result of a single exposure to a substance; in contrast to chronic.

acute bronchitis disease that presents with cough productive of sputum that is often secondary to a viral or bacterial infection and the inhalation route of exposure to high concentrations of bioaerosols in agricultural facilities.

acute reaction characterized by a relatively short time period between exposure and effect; generally involves a dramatic onset of an adverse health effect.

acute toxicity characterized by a single dose or single exposure to a substance that results in an adverse impact.

***N*-acyl homoserine lactone (AHL)** small, diffusible signal molecules used by gram-negative bacteria for intercellular communication.

adenine a purine base that is one of the four nucleotides comprising DNA and RNA molecules; adenine forms a bond with thymine on the opposite strand of the DNA molecule.

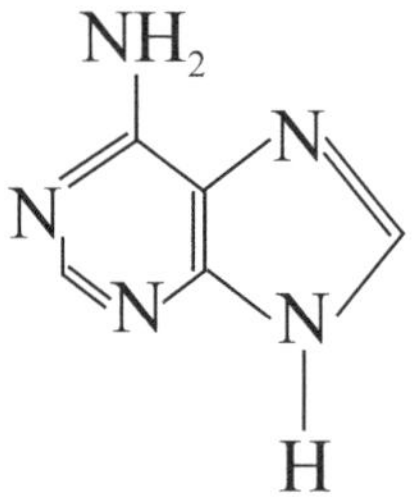

adenosine a nucleotide ($C_{10}H_{13}N_5O_4$) composed of adenine and ribose that is found in RNA.

adenosine triphosphate (ATP) a carrier composed of adenosine and three phosphate groups that transports phosphate and energy in biological systems; uniformly present in all microorganisms and the concentration is relative to cell carbon but dependent on the physiological state of the cell and lost rapidly with cell death; used as a measurement of microbial biomass; detected by the luciferin-luciferase assay and high performance liquid chromatography.

adenovirus member of the Adenoviridae family; any one of a group of double-stranded DNA viruses that have an icosahedral capsid, approximately 60–90 nm in diameter; cause acute respiratory and ocular infections; some cause mild gastrointestinal illness; spread by direct contact, fecal-oral route of transmission, and waterborne transmission unusually stable at extreme pH conditions and are highly resistant to chemical or physical agents, allowing for prolonged survival outside of the body.

adherence the binding of a microorganism to a surface.

adhesins macromolecules on the cell surface or appendages that make it easier for cells to adhere to other cells or a surface material.

adhesion binding to a surface.

adhesion site the region of a cell that is specialized for adhesion.

adjuvant a substance that enhances the action of a reagent; a substance the increases the antigenic response of another substance.

adnate broadly attached.

adsorption the adherence of material to the outer layers of another substance; in contrast to absorption.

adsorption coefficient a defined constant used in chromatography relating the binding of a molecule to the stationary phase; also use to characterize binding of a microorganism to a solid surface such as a soil particle.

adventitious septa cellular division observed in some fungi that is formed in the absence of nuclear division, generally in association with the movement of cytoplasm from one part of the organism to another.

aerated pile composting a method of composting in which air or oxygen is injected into the material rather than turning of the pile.

aerial hyphae hyphal structures that project upward from the surface of a fungal colony.

aerial mycelia masses of hyphae that project upward from the surface of a fungal colony.

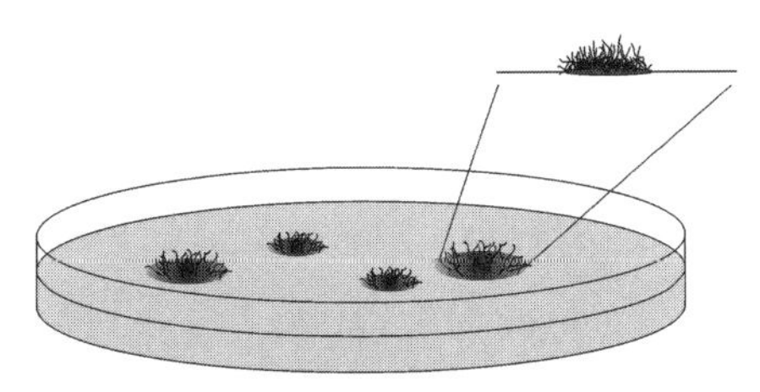

aerobe/aerobic a microorganism that grows in the presence of oxygen; subdivided into facultative, obligate, and microaerophilic classifications.

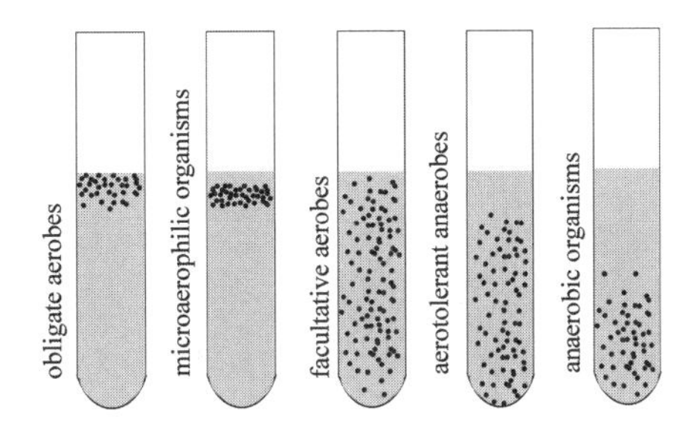

aerobiology the study of bioaerosols including airborne bacteria, fungal spores, fragments of microorganisms, and by-products of microbial metabolism.

aerodynamic particle size physical measurement of an airborne particle that incorporates size, shape, and density.

aerodynamic particle sizer (APS) laser-based instrument used to measure the airborne concentration of particles within a defined size range; can be used to estimate the concentration of bioaerosols without constraints of viability or physiological state.

aerogenic gas-producing; in contrast to anaerogenic.

Aeromonas bacterial genus; member of the Vibrionaceae; gram-negative, chemoorganotrophic, straight bacilli with rounded ends that are motile with a single polar flagellum although there is one recognized species that is nonmotile; isolated from fresh water and sewage.

Aeronomas salmonicida bacterial species; causative agent of fish disease, commonly occurring in hatcheries.

aerosol fine suspension of liquid or particulate in the air.

aerotolerant an anaerobic microorganism that can survive in the presence of oxygen.

AES auger electron spectroscopy.

afferent leaning toward; in contrast to efferent.

affinity the strength of interaction between two entities.

affinity chromatography analytical technique in which the stationary phase (e.g., a substituted agarose, cellulose, dextran, or polyacrylamide) interacts with the analyte in a sample because of a biochemical reaction or chemical affinity, generally used to describe antigen/antibody reactions or enzymatic reactions.

aflatoxin toxic metabolite produced by *Aspergillus flavus*, *Aspergillus parasiticus*, and *Aspergillus nomius*; classified by the International Agency for Research on Cancer (IARC) as having sufficient evidence for human and animal carcinogenicity, and may be involved in occupational respiratory cancers among food and grain workers.

aflatoxin B_1 mycotoxin referred to as the most carcinogenic naturally occurring substance known.

AFM atomic force microscopy.

AFR accidental fecal release.

agar a polysaccharide extract of algae used to solidify reagents into a semisolid matrix for the growth and isolation of microorganisms.

Agaricales fungal order; includes basidiomycetes.

agarose a polymer purified from agar that forms a gel when cooled; used to prepare a support matrix for separation of macromolecules using gel electrophoresis; also used as a support in affinity chromatography.

agglutination rapid formation of a matrix of cells or particles following the addition of specific reagents.

agglutinin a reagent that induces agglutination.

aggregation the formation of a matrix of cells or particles that is slower than agglutination because not every particle or cell participates in the matrix.

AGI-30 sampler all-glass impinger sampler.

agitoxin peptide that inhibits potassium channels.

Agrobacterium bacterial genus; gram-negative bacilli with flagella that produce tumorous growths, termed galls, on infected plants.

Agrobacterium rhizogenes bacterial species; phytopathogen; causative agent of hairy root of apples.

Agrobacterium rubi bacterial species; phytopathogen; causative agent of cane gall of raspberries.

Agrobacterium tumefaciens bacterial species; phytopathogen that infects many different plants producing a tumor-like growth termed crown gall.

AHL *N*-acyl homoserine lactone.

A horizon mineral region of the soil that lies near the surface and is characterized as a zone of maximal leaching; generally divided into an A_1 horizon where the mineral soil is mixed with humus, an A_2 horizon which is an area of maximal leaching of silicate clays, iron oxides, and aluminum oxides, and an A_3 horizon that is a transition region above the B horizon.

AIHA American Industrial Hygiene Association.

airborne present in the air; transported through the air.

Air-O-Cell a commercially available plastic, individual cassette used as an impactor sampler with a slit inlet designed for the collection of airborne fungi and pollen by impaction sampling; collection of samples is achieved with a vacuum pump operated at 15 liters/min.

air pollution the presence of contaminant material in the air; generally limited to abiotic contaminants such as particulate, volatile organic compounds, and by-products of combustion, but may also be used to describe the presence of biotic contaminants in the air.

alasan the bioemulsifier produced by *Acinetobacter radioresistens*.

alcohol fermentation the conversion of glucose to ethanol; process used in wine making in which the glucose from crushed grapes and juice is converted to ethanol, CO_2, and heat by the action of *Saccharomyces cerevisiae*.

alcohol lamp small, compact glass container fitted with a wick and supplied with ethanol as fuel for use in flame sterilization.

aldrin hexachloro-hexahydrodimethanonaphthalene; a pesticide that is resistant to biodegradation; in contrast to carbaryl.

aleurioconidium/aleuriospore a fungal spore that is produced asexually by the swelling of a terminal or lateral cell of a hypha.

Alexander, Martin 20th century environmental microbiologist who formalized the principle of microbial fallibility in 1965.

algae photosynthetic eukaryotic organisms that are separated from plants due to their lack of tissue differentiation; function in aquatic systems as primary producers; associated with allergic reactions with the dermal route of exposure and the inhalation route of exposure; cells are found as unicellular or colonial (aggregates of cells), some are linked end to end in a filamentous form that may be branched or unbranched; in contrast to cyanobacteria.

aliquot a small portion of a sample or reagent.

alkalinity the capacity of a substance to combine with hydrogen ions; generally expressed as equivalents of carbonate or bicarbonate ions.

alkaliphile/alkaliphilic a microorganism that grows best at a high pH (>8.0).

alkaloids organic nitrogenous bases; generally crystals but maybe liquid or gelatinous; some are insecticidal and some mycotoxins belong to this group of compounds.

alkyl benzyl sulfonates (ABS) chemical compounds with a polar sulfonate group and a non-polar alkyl end that are components of anionic detergents, linear structures can be degraded in nature but nonlinear structures are resistant to biodegradation when they enter the environment and cause foaming of rivers and streams.

allantoid sausage-shaped.

Allee's principle both positive and negative interactions of organisms are interrelated and population dependent and they may occur even within a single population.

allelopathic the production of a substance by one microorganism that adversely affects another organism.

allergen a compound primarily composed of protein that acts as an antigen and elicits an allergic response.

Allergenco a spore trap sampler that utilizes forced air flow sampling with a slit inlet designed for the collection of airborne fungal spores and pollen onto a sticky surface preparation on a glass microscope slide; time discrimination results from the movement of the glass slide under the slit.

allergenic reaction an immune response resulting from exposure to an allergen by a sensitized individual.

allergic alveolitis inflammation of the alveoli of the lung caused by inhalation of organic particulate into the terminal airways; associated with inhalation of allergenic fungal spores.

allergic bronchopulmonary aspergillosis (ABPA) a pulmonary disease resulting from an *Aspergillus fumigatus* infection of the lower respiratory tract with the growing mycelium of the fungus continually releasing antigen and mycotoxin resulting in allergenic inflammation and an immunotoxic effect.

allergic fungal sinusitis (AFS) a chronic disease observed in approximately 7% of chronic sinusitis patients; a dark gray-black thick accumulation of material present in the sinuses that obstructs drainage and provides an environment for bacterial sinusitis; may be the result of allergenic inflammation and the immunotoxic effect of fungal metabolites, especially species of *Aspergillus* and *Fusarium*, and the dermatiaceous fungi.

allergic response / allergy an immunological reaction to an antigen.

allergic rhinitis genetically determined, IgE-mediated responses of nasal stuffiness and drainage, and sinus headaches following inhalation exposure to grasses, trees and some fungi.

all-glass impinger sampler (AGI-30) a glass impingement sampler designed with a curved inlet for the collection of bioaerosols into a liquid collection medium using a vacuum pump operating at 12 liters/min.; the distance from the jet to the bottom of the collection tube is 30 mm.; other AGI samplers utilize other distances (e.g., 4 mm for the AGI-4) but decreased distance increases sampling stress on biological particles.

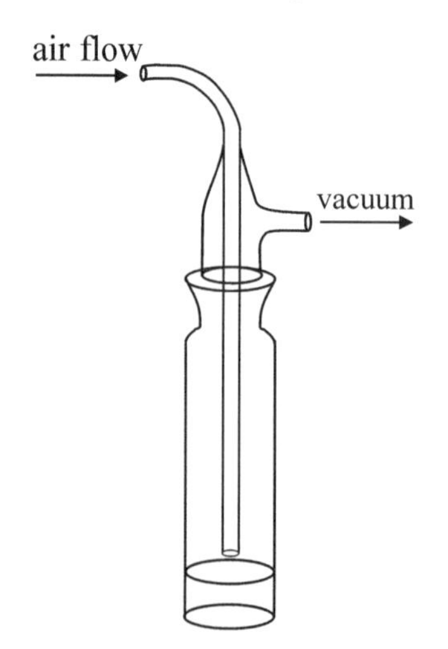

allochthonous transient members of a habitat; organisms that are foreign to an environment and do not occupy a functional niche.

allogenic succession the alteration of a habitat by environmental factors that permits new populations to develop; in contrast to autogenic succession.

alopecia baldness, hair loss.

alpha error Type I error.

Alt a I major allergen present in *Alternaria*; some evidence that it is the major allergen in *Stemphylium* and *Ulocladium*.

Alternaria fungal genus; floccose to velvety colony on malt extract agar that is pale gray to olive or dark green to brown in color with a brown to black reverse; characterized by large multicellular, septate, muriform, ovoid or obclavate conidia that have a beak-like apical cell that help to distinguish this fungus from *Ulocladium*; conidia are produced in chains but may appear singly; ubiquitous and common worldwide; approximately 40–50 species; a plant pathogen of drought and plants injured by insects, notably leaf spot and blights, and is used as a biological control agent of weeds; isolated from soil, organic debris, food stuffs, textiles, and indoors with a water activity of 0.85–0.88; associated with Type I allergies and Type III hypersensitivity; major allergen Alt a I.

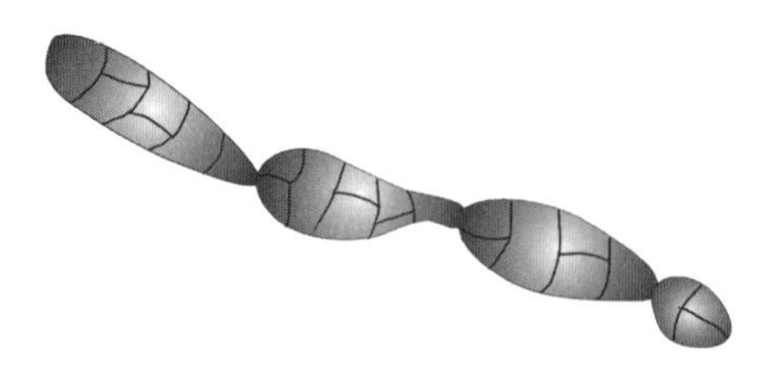

Alternaria alternata fungal species; widespread saprophyte, mycotoxin producer, allergen, and phytopathogen.

Alternaria porri fungal species; phytopathogen of onion.

Alternaria solani fungal species; phytopathogen of potato and tomato.

alternating arthroconidia the placement of arthroconidia between vegetative cells in a chain such that when the vegetative cells degenerate and break apart the spores are released to the environment; characteristic feature of *Coccidioides*; in contrast to simple arthroconidia.

alveolar macrophage defects changes in macrophage cells of the lung that can be caused by exposure to trichothecene mycotoxins.

alveolate pitted, like a honeycomb.

AM arithmetic mean.

amalgam an artificial consortium of organisms.

Amanita fungal genus; member of the family Amanitaceae within the basidiomycetes; characterized by free gills and the presence of an annulus and a volva; several species are poisonous.

Amanita phalloides fungal species; known as the death cap due to its association with numerous deaths following ingestion.

Amastigomycota a primary taxonomic grouping of fungi that do not produce motile cells; the four subdivisions of this grouping include the Ascomycotina, the Basidiomycotina, the Deutromycotina, and the Zygomycotina.

amensalism/amensalistic relationship of microbial populations in which one population produces a substance that is inhibitory to the other population; also termed antagonism.

American Conference of Governmental Industrial Hygienists (ACGIH) a member-based organization of more than 4,000 professionals worldwide focused on industrial hygiene issues; members provide information to government, corporations, and academia; located in Cincinnati, Ohio.

American Industrial Hygiene Association (AIHA) a professional organization that promotes industrial hygiene and occupational health issues for the health and well being of workers, the community, and the environment.

American Society for Microbiology (ASM) a life science organization started in 1899 representing a variety of disciplines within the field of microbiology; based in Washington, DC currently with 42,000 members worldwide.

American Society of Testing Materials (ASTM) ASTM International.

American Type Culture Collection (ATCC) major culture collection entity located in Manassas, VA, USA that catalogs and sells standard strains of bacteria, viruses, cells, and fungi.

American Water Works Association (AWWA) an international nonprofit scientific and educational society founded in 1881 that is dedicated to the improvement of drinking water quality and supply.

American Water Works Association Research Foundation (AWWARF) the research component of the American Water Works Association that serves as a vehicle for the drinking water community to collectively underwrite a centralized research effort on relevant issues; most of the funding for the research program is provided by voluntary subscriptions by water utilities.

Ames test a bacterial assay that utilizes specific mutant strains of *Salmonella typhimurium* to determine the potential toxicity and mutagenicity of chemicals.

AMF arbuscular mycorrhizal fungi.

amoeba, amoebae protozoa with environmentally resistant cysts that survive in soils and water and trophozoites that are motile by cytoplasmic streaming.

amplicon a fragment of DNA that undergoes amplification.

amplified ribosomal DNA restriction analysis (ARDRA) a technique in molecular biology in which the 16S rRNA gene (ribosomal DNA) is amplified, and the resulting amplicon is digested using restriction enzymes to obtain a profile that can be used to identify microorganisms by comparing the profile to a library of profiles from known organisms.

amplify/amplification increase in the number of something, such as the increase in the number of copies of a sequence of DNA in polymerase chain reaction amplification.

amphiphilic having a polar and nonpolar region.

amphitrophic an organism that can function as a chemotroph or as a phototroph.

amphoteric having both an acidic and a basic component.

anabolic/anabolism the biochemical processes that involve the synthesis of cell constituents from simpler molecules, usually requiring energy; in contrast to catabolism.

anaerobe/anaerobic a microorganism that grows in the absence of oxygen.

anaerobic digester two-step wastewater treatment process used to convert odorous sludge to an inert non-odorous product under anaerobic conditions; during the first step complex organic materials are depolymerized and converted to CO_2, H_2, and fatty acids by the action of various facultative and obligate anaerobic bacteria followed by the generation of methane by methanogenic bacteria.

anaerogenic non-gas-producing; in contrast to aerogenic.

analysis of covariance (ANCOVA) statistical analysis used in prospective studies where two groups differ on one or more variables that could be related to the dependent variable.

analysis of variance (ANOVA) statistical analysis used to determine if a difference exists between groups when more than two groups are studied.

analyte the subject of the analysis.

anamorph term used in mycology to describe the asexual state.

anammox anoxic ammonia oxidation.

anastomose forming a network.

anchorage attachment without adhesion.

anchorage dependent requirement that a cell be attached to a surface for growth.

ANCOVA analysis of covariance.

Andersen impactor sampler an impactor sampler developed in 1958 by A. Andersen for collection of airborne bacterial cells and fungal spores that are operated by an external vacuum pump operating at a fixed air flow of 28.3 liters/min.; single (Andersen N6), two-stage, and six-staged varieties are commercially available for use with follow-on culture analysis with collection of airborne particles through a sampling cone and directed through stainless steel plate(s) containing precision drilled holes of a specified diameter and subsequent impaction onto an agar-filled petri plate; positive-hole correction recommended to minimize enumeration errors due to the presence of multiple particles passing through the same hole and landing on to the same area of the agar surface.

Andersen N6 Andersen single-stage impactor sampler.

Andersen single-stage impactor sampler single stage Andersen impactor sampler consisting of the 6th stage of the multi-stage version for the collection of particles ≤10 μm in diameter.

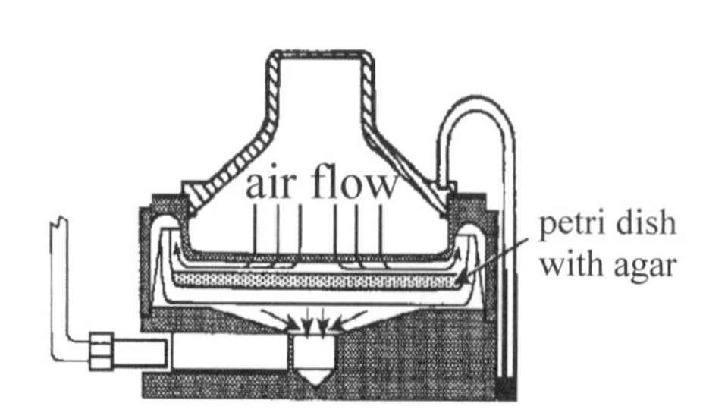

anergy/anergic the absence of a reaction to antigens or antibodies.

Ångstrom unit a unit of measure; 10^{-10} m.

animal studies experiments involving animal subjects.

animal virus a virus that infects animals; in contrast to bacteriophages and plant viruses.

anisotropic not the same in all directions.

anneal hybridization of DNA following dissociation by heat.

annealing temperature the temperature at which two complementary strands of nucleic acid will reassociate after temperature-induced denaturation or "melting"

annellide a conidiogenous cell with a slight elongation and ringed with a succession of circular scars from the production of each conidium.

annulate having a ring-like structure.

annulus a ring-like structure.

ANOVA analysis of variance.

anoxia / anoxic total lack of oxygen.

anoxygenic the absence of oxygen production.

antagonism amensalism.

anthrax disease resulting from inhalation or dermal exposure to *Bacillus anthracis*.

anthrax lethal toxin (LeTx) the toxin that produces shock-like effects in mammals infected with *Bacillus anthracis*.

anthrax edema factor a toxin produced by *Bacillus anthracis*.

anthropogenic man-made.

antibacterial a bacteriostatic or bactericidal substance.

antibiotic an antimicrobial substance administered to combat a bacterial infection or used as an amendment in culture medium to permit the growth of select organisms and prevent the growth of other microorganisms.

antibiotic amendment addition of a broad spectrum antibiotic or specific antibacterial agent to a culture medium to minimize the growth of unwanted bacteria that may be present in the sample.

antibiotic resistance the acquired ability of a microorganism to tolerate an antimicrobial substance.

antibody an immunoglobulin that interacts with the antigen that induced its formation.

antichaotropic a substance that maintains the stability of the structure of water by decreasing the solubility of nonpolar substances in a polar solvent (such as citrate or magnesium sulfate in water) and decreases the solubility of viruses in solution; substance that increases hydrophobic interactions in water resulting in the adsorption of some viruses to surfaces.

antifungal a fungistatic or fungicidal substance that can combat a fungal infection or used as an amendment in culture medium or other materials to prevent the growth of fungi and permit the growth of other microorganisms.

antifungal amendment addition of an antifungal agent to a culture medium to minimize growth of fungi that may be present in the sample.

antigen a substance that induces an immunogenic response and interacts with a T cell receptor or an immunoglobulin.

antigenic drift minor changes in viral proteins resulting from mutation; in contrast to antigenic shift.

antigenic shift major changes in viral proteins due to reassortment of genes; in contrast to antigenic drift.

antimicrobial harmful to microorganisms.

antiseptic a chemical that kills or inhibits microorganisms but is not harmful to human tissue.

antiviral a viricidal substance.

AO acridine orange.

AOB autotrophic ammonia-oxidizing bacteria.

apex tip.

aphotic zone the area of an aquatic habitat in which insufficient light penetrates, preventing photosynthesis from occurring.

API a commercially-available identification kit that is based on specific enzymatic reactions by microorganisms.

apical the farthest point from the base.

apices at the tips.

apiculate ending as a short pointed tip.

apnea temporary cessation of breathing.

apochromatic lens an objective lens that corrects chromatic aberration for three colors and spherical aberration for two colors resulting in the production of high quality images revealing true colors without distortion of shape.

apophysis funnel-shaped swelling.

apoptosis programmed cell death.

apothecium cup-shaped or saucer-shaped fruiting body that contains asci.

appressed flattened.

APS aerodynamic particle sizer.

AR analytic reagent grade designation for a chemical.

arbovirus a virus transmitted by an arthropod (e.g., mosquitoes and other insects, arachnids, crustaceans).

arbuscular mycorrhizal fungi (AMF) fungi with finely branched hyphae that penetrate host tissue.

arachnoid cobweb-like.

Archaea a phylogenetic grouping at the domain level of microorganisms that is distinctly different from the prokaryotes in the domain Bacteria and the eukaryotes in the domain Eukarya; chemotrophic, some are chemolithotrophic, and many are thermophilic.

Archaeoglobus bacterial genus; characterized as irregular, motile spheres of group I non-acetate oxidizing dissimilatory sulfate reducing bacteria of the Archaea that are hyperthermophiles with a temperature growth range of 64–92°C and an optimum growth temperature of 83°C.

archival/archiving storage of specimens for future analysis.

arcuate moderately curved.

ARD acid rock drainage.

ARDRA amplified ribosomal DNA restriction analysis.

area method a means of refuse disposal in a sanitary landfill in which waste is spread and compacted over a large surface and covered with a layer of soil at the end of the day's operation; in contrast to the trench method and the ramp variation.

ARISA automated rRNA intergenic spacer analysis.

arithmetic mean (AM, $\bar{X}$) the average of a set of values; the measure of the typical value or central tendency for interval and ratio data.

$$\bar{X} = \frac{\Sigma X}{N}$$

Arthrinium fungal genus; cottony to wooly colony on malt extract agar that is white in color with brown spots on the surface and pale on the reverse; the hyphae are septate and hyaline with pale, short or elongate obclavate conidiogenous cells that form dense clusters of brown, lens-shaped conidia with an equatorial slit; isolated from soil and decomposing plant material.

arthroconidia fungal spores formed by the disarticulation of hyphal cells and liberated by fission or lysis.; alternating barrel-shaped spores are a diagnostic structure in cultures of *Coccidioides immitis*.

articulate jointed.

asci plural of ascus.

ascigenous having an ascus.

ascitic fluid serous fluid in the peritoneal cavity.

ascocarp former terminology for ascomata.

ascomata spherical or flask-shaped fruiting structure containing asci.

Ascomycetes Ascomycota.

Ascomycota fungal group; largest grouping of fungi that is characterized by the presence of an ascus with reproduction through the production of ascospores that frequently are observed in spore trap air samples; includes saprophytes, phytopathogens, and lichen-forming members; commonly found in soil and decaying plant material.

Ascomycotina a subdivision of the Amastigomycota; grouping of saprophytic, symbiotic, or parasitic fungi that are characterized as unicellular or having septate mycelium and producing ascospores in asci.

ascospore reproductive structure produced in an ascus produced by free-cell formation; characteristic of more than 3,000 fungal genera that are ubiquitous in nature; forcibly discharged during periods of high humidity or rain; potential for pathogenicity or toxin production dependent on genus and species.

ascus a sac-like cell that produces ascospores and is characteristic of the Ascomycotina.

aseptate without crosswalls.

aseptic/aseptic technique sterile conditions; absence of microbial contamination.

ASHRAE American Society of Heating, Refrigerating, and Air Conditioning Engineers.

ASM American Society for Microbiology.

aspergillosis respiratory disease caused by the inhalation of spores of some species of *Aspergillus*.

Aspergillus fungal genus; member of the Deuteromycotina; ubiquitous, saprophytic organism common to cultivated soils, decomposing plant material, and stored grain, and is the causative agent of rot of seeds; hyaline, septate hyphae; conidiophore terminating in an apical vesicle housing phialides; distinctive basal foot cell; culture-based identification based on colony color and the form of the conidial head; approximately

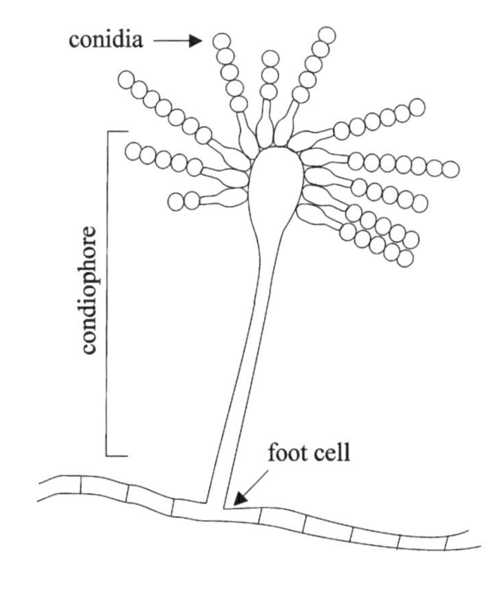

200 species; water activity of 0.71–0.94 dependent on species.

Aspergillus flavus fungal species; colony on malt extract agar is velvety in texture, yellow to green or brown in color with golden to red-brown reverse; conidiophores are variable in length with a rough, pitted, or spiny surface; uniseriate and biseriate phialides cover the entire vesicle and radiate in all directions; aflatoxin producer.

Aspergillus fumigatus fungal species; thermotolerant organism with an optimal growth temperature ≥40°C; often isolated at 48°C with confirmation by culture on potato glucose agar or cornmeal agar at 37°C; colony on malt extract agar is velvety, downy, or powdery in texture, initially white in color turning generally smokey gray-green or blue-green with white, tan or yellow reverse; some isolates may also show a lavender diffusable pigment in the surrounding agar; conidiophores are short with smooth walls; uniseriate phialides usually on upper 2/3 of vesicle in parallel to axis with the conidiophore forming a columnar head; commonly isolated from soil, plant material and composting facilities; opportunistic pathogen; inhalation exposure may result in Type I allergies and Type III hypersensitivity.

***Aspergillus glaucus* group** grouping of fungal species; colony on malt extract agar is felt-like or downy/powdery in texture, green in color with yellow areas, occasionally brown with yellow to maroon or chestnut brown reverse; conidiophores are variable in length with smooth walls; uniseriate phialides, radiate to loosely columnar covering the entire vesicle; prominent yellow cleistothecia generally present; produce ascomata in a sexual stage containing pale ascospores with or without equatorial crests; commonly isolated from soil, plants, and house dust; osmophilic preferring culture medium with 20% sucrose; some species are xerophilic.

Aspergillus nidulans fungal species; colony on malt extract agar is dark green with orange to yellow areas containing cleistothecia with a purple reverse and brown to purple exudate; hülle cells present; short, columnar conidial heads with sinuous, smooth-walled conidiophores with shades of cinnamon brown; important fungus in the production of penicillin; produces sterigmatocystin, a mycotoxin.

Aspergillus niger fungal species; colony on malt extract agar is wooly in texture, initially white to yellow turning black with white to yellow reverse; conidiophores are long with smooth walls; biseriate phialides covering the entire vesicle forming a radiate head; brown conidia with warts, spines, or ridges; commonly isolated from house dust, soil, textiles, plant litter, seeds, fruits, and dried nuts; industrially used to degrade wastes from food production such as apple juice, potato, and sugar beet facilities; spoilage organism in meat processing and onion storage; causative agent of allergic airway disease, especially in industrial settings.

Aspergillus ochraceus fungal species; colonies are zonate and buff in color ranging to yellow with a yellow to greenish brown reverse; thick walled, rough conidiophores are pigmented a dull yellow to light brown; conidia are globose; young sclerotia are white to pink turning lavender to purple; primarily isolated in subtropical and tropical climates, but may be found in house dust, soil following the land application of sewage sludge; spoilage organisms of tobacco, leather, wood pulp, and cotton, but is also used for in biological control of some phytopathogens and in the degradation of *n*-alkanes; exposure of farmers; production of ochratoxin under laboratory conditions, but not found in feed or grain contaminated with this organism (see *Penicillium verrucosum*).

Aspergillus parasiticus fungal species; compact yellow, grass green to cedar green colony on malt extract agar with a uncolored to cream reverse; colorless, variable length conidiophores that are smooth or roughened with globose echinulate bright yellow conidia; aflatoxin producer when growing in sugar cane and groundnuts.

Aspergillus terreus fungal species; colony on malt extract agar is velvety in texture, cinnamon brown with white to brown reverse; conidiophores are short with smooth walls; biseriate phialides, compactly columnar; round hyaline cells are produced on mycelia submerged in agar.

Aspergillus versicolor fungal species; colony on malt extract agar is downy to powdery in texture and white, yellow, beige to orange-yellow, yellow-green or emerald green with a pale or yellow in color with a orange to purple reverse, a clear to wine-red exudate may appear on the agar surface; radiate conidial heads; conidiophores are >300 μm in length with smooth walls; biseriate phialides that are solitary or grouped in a brush-like cluster; hülle cells sometimes present; isolated from soil, plant materials, foodstuffs (especially cheese), and indoors in buildings with water intrusion problems on building materials, painted surfaces, wallpaper, carpeting, and in house dust; may produce a carcinogenic mycotoxin (sterigmatocystin) and ethylhexanol, a volatile organic compound that causes a strong moldy odor depending on the substrate.

asperulate slightly roughened.

asporagenous without spores.

assay as a noun it is an analysis; used as a verb it refers to conducting an analysis.

assimilative/assimiliatory sulfur reduction conversion of SO_4^{2-} to organic sulfur for use in biosynthesis such as formation of amino acids; in contrast to dissimilative/dissimilatory sulfur reduction.

asthma a reversible airway obstructive inflammatory disease resulting from an allergenic response that is usually mediated by IgE; genetic susceptibility of the affected individual and environmental exposure are involved with increased incidence and increased severity associated with indoor allergens; causal relationship to exposure to house dust mite, cockroach, and cat dander; an association with exposure to endotoxin, fungi, dogs, and rhinoviruses.

asthma-like syndrome symptoms of chest tightness, shortness of breath, and dry cough in the absence of an allergic process.

ASTM International a not-for-profit international organization founded in 1898 with over 30,000 members of producers, users, consumers, and government and academia representatives from 100 nations; provides standards for manufacturing, procurement and regulatory activities that are accepted and used in research and development and product testing; formerly known as the American Society for Testing and Materials (ASTM).

astrovirus genus of small (27–30 nm diameter), round, single-stranded RNA viruses in the family Astroviridae; found in fecal material; transmitted by the fecal-oral route of transmission causing gastroenteritis.

asymptomatic lack of identifiable signs or symptoms.

asymptotic of or pertaining to a line whose distance to a given curve approaches zero.

ATAD Atmospheric Transport and Dispersion model.

ataxia loss of coordination or motor skills.

ATCC American Type Culture Collection.

atmospheric pressure the force exerted by the weight of the air column at the surface of the earth which at sea level is 1 atmosphere or 760 mm mercury; in contrast to hydrostatic pressure.

Atmospheric Transport and Dispersion (ATAD) model computer-based program designed to calculate the trajectory of airborne fungal spores of agricultural importance and provide information on the concentration of the spores and deposition pattern.

atomic force microscopy (AFM) a microscopic technique that utilizes a probe and piezoelectric sensors to measure surface characteristics at the atomic level providing three-dimensional imaging.

atopic/atopy genetic predisposition of allergen sensitivity.

ATP adenosine triphosphate.

ATR acid tolerance response.

atrium cavity.

attack rate the number of individuals in an exposed population that show signs of illness during a defined, limited period of time, such as an epidemic.

attenuate to taper off or draw-out; to lessen the virulence of a pathogen.

attributable risk the amount or proportion of disease that can be associated with a specific exposure.

auger electron spectroscopy (AES) an analytical technique used to map the elemental composition of surfaces.

auramine O stain a yellow to orange reagent that is used as a dye for microscopic analysis.

Aureobasidium fungal genus; yeast-like colony, beginning cream to pink becoming dark brown in color on malt extract agar; ubiquitous, organism found in soil, forest soils, fresh water and marine estuary sediments, plants, fruit, wood, and indoors in areas with an accumulation of moisture such as kitchens and bathrooms, tile grout, and window sills.

Aureobasidium pullulans fungal species; associated with Type I allergies, Type III hypersensitivity, and hypersensitivity pneumonitis; isolated from stored wet cardboard; produces a biodegradable polysaccharide with shiny rayon-like appearance and nylon-like strength and also used to remove undesirable components of raw textile materials; formerly termed *Pullularia pullans*.

autecology study of ecological systems that focuses on life history and behavior of individual populations as a means of adaptation to their environment; in contrast to synecology.

autochthonous indigenous or naturally occurring, physiologically compatible with the physical and chemical environment.

autoclave/autoclaving instrument/procedure used for sterilization of laboratory supplies, reagents, and waste materials that utilizes heat and pressurization; chamber temperature typically maintained at 121°C with 15 psi for 15 minutes for sterilization of liquids, 30 minutes for solids, and 60 minutes for waste materials.

autoclave tape adhesive material that changes appearance or highlights letters to indicate exposure to heat, but does not validate the proper operation of an autoclave as indicated by the use of a spore strip.

autofluorescence excitation of light that is exhibited by chloroplast material in algae and may occur with some materials in a sample matrix resulting in interference with direct count methods of detection.

autogenic succession the modification of a habitat by a microbial community that permits new populations to develop; in contrast to allogenic succession.

autoinducer a signal protein secreted by microorganisms involved in cell to cell communication to influence the behavior of other members of the microbial population.

autologous of self.

autolysis the degradation of a cell that is caused by the cell itself.

automated DNA sequencer a machine that is used to determine the sequence of nucleotide bases in a piece of DNA.

automated rRNA intergenic spacer analysis (ARISA) a system that uses a fluorescent-tagged oligonucleotide primer with an automated polymerase chain reaction amplification and electrophoresis detection.

autophagy removal of membrane organelles and other components in the cytoplasm by the cell's own enzymes.

autoradiography method for the detection of radioactive particles; used for the detection of nucleic acids using radioactive phosphate ($^{32}PO_4$) added to a sample during nucleic acid synthesis or proteins using radioactive sulfate ($^{35}SO_4$) added during growth.

autosampler mechanized means to repeatedly withdraw an aliquot from a sample matrix.

autotroph a microorganism that is able to utilize CO_2 or bicarbonate as a sole carbon source.

autotrophic the synthesis of organic matter from inorganic sources.

autotrophic ammonia-oxidizing bacteria (AOB) bacteria that participate in the nitrogen cycle by oxidizing NH_4^+ to NO_2^- in the environment.

autotrophic succession occurs in a community of pioneer organisms when there is a minimum of organic matter and an unlimited supply of solar energy.

autoxidation the oxidation of a substance resulting from exposure to the air.

auxesis growth by increasing the size of the cell rather than the number of cells.

auxotroph a microorganism that has acquired a nutritional requirement by genetic mutation.

avellaneous descriptive of a drab to hazel color.

avidity strength of binding.

Aw, a_w water activity.

AWWA American Water Works Association.

AWWARF American Water Works Association Research Foundation.

axenic when all the biological populations are defined or known; also termed gnotobiotic.

Azomonas bacterial genus; free-living nitrogen fixing bacilli that are primarily aquatic; unlike *Azotobacter* they do not produce resting cysts.

Azospirillum bacterial genus; free-living nitrogen fixing spirillum-shaped bacteria that are symbionts with plants, especially corn.

Azotobacteraceae bacterial family; characterized by their ability to fix molecular nitrogen.

Azotobacter bacterial genus; member of the Azotobacteraceae; gram-positive, obligate aerobic nitrogen fixing free-living bacilli; capable of forming resting structures called cysts that are not completely dormant or heat resistant like endospores, but demonstrate resistance to desiccation, ultraviolet light and ionizing radiation.

B

β Greek symbol for beta that is used to denote the probability that a statistical Type II error will be made.

bacillate rod-shaped.

bacilli plural of bacillus.

bacillus descriptive of a rod-shaped bacterium; may be slender or broad with tapered or blunted ends depending on genus.

Bacillus bacterial genus; gram-positive, aerobic or facultatively anaerobic, endospore-forming, chemoorganotrophic bacilli with peritrichous flagella; ubiquitous in soil; increased resistance to desiccation, ultraviolet light.

Bacillus anthracis bacterial species; causative agent of anthrax; generally the result of inhalation or dermal contact and is commonly associated with animal handlers, veterinarians, rug manufacturing using animal fur; possible use in bioterrorism or biological warfare as a weapon of mass destruction.

Bacillus cereus bacterial species; associated with both diarrheal food poisoning and emetic food poisoning; endospores survive normal cooking procedures and germinate during improper storage after cooking.

Bacillus licheniformis bacterial species; opaque colony with a dull to rough surface and hair-like outgrowths; colony is firmly attached to the agar surface and aged cultures may become brown in color; bacilli are often in chains; endospores occur in soil and may survive extreme heat; produces bacitracin.

Bacillus stearothermophilus bacterial species; thermophile with capability of growth at 65° occurs in soil and hot springs, desert sand, compost, food, and in ocean sediment and arctic waters.

Bacillus subtilis bacterial species; ubiquitous mesophile; commonly used as a surrogate for pathogenic *Bacillus* spp. in laboratory experiments.

***Bacillus thuringiensis* (*Bt*)** bacterial species; ubiquitous leaf saprophyte, soil microorganism, and insect pathogen that can naturally produce crystalline proteins with insecticidal properties as a result of gene sequences present on plasmids; currently more than 1000 strains, 100 of which have insecticidal protein sequences; produces Cry proteins and Cyt proteins that are insecticidal and dried preparations are used as a microbial pest control agent marketed since the 1960s for insect control with more than 400 preparations currently registered; isolated from soil, dust in stored grain, insects, and coniferous and deciduous trees; first discovered in diseased silkworms in 1901.

bacitracin peptides produced by *Bacillus licheniformis* that inhibit synthesis in gram-positive bacteria.

background risk likelihood of contracting a specific disease without exposure to the known causative agent.

bacteria ubiquitous, single celled prokaryotic microorganisms; the domain level of taxonomy of microorganisms that are distinctly different from prokaryotes in the domain Archaea and the eukaryotes in the domain Eukarya.

bacterial lawn confluent, uniform distribution of bacterial cells across the surface of an agar medium.

bactericidal/bacteriocide an agent capable of killing bacteria.

bacteriochlorophylls chlorophyll pigments present in bacteria.

bacteriocin an antibiotic-like substance that is produced by bacteria against other closely related bacteria.

bacteriology the study of bacteria and members of the Archaea.

bacteriophage virus that infects prokaryotic cells.

bacteriorhodopsin retinal-containing, membrane-bound protein produced by some halophiles capable of light-mediated proton motive force formation.

bacteriostatic capable of inhibiting the growth of, but not killing bacterial cells.

bacterium singular term of bacteria.

bacteroid a *Rhizobium* cell capable of nitrogen fixation that is found in a root nodule characterized as swollen, branched or misshapen.

Bacteroides bacterial genus; obligate anaerobes characterized as non-spore forming, gram-negative, saccharolytic commensals found in the intestinal tracts of humans and some other animals; several species isolated from the rumen have been reclassified as *Fibrobacter* or *Prevotella* spp.

baculovirus virus; a DNA virus isolated from insects; pathogenic to insects, but has a limited host range.

ballistoconidia/ballistospore a spore that is actively discharged from a basidium most often the result of high moisture or water droplet contact.

ballistospore colorless basidiospore produced by some fungi.

BAPP test biological acid-producing potential test.

bar chart/bar graph presentation of data using the length of a bar to denote value; similar to a histogram plot except that there are spaces between the bars; a dot plot can be used as a variation.

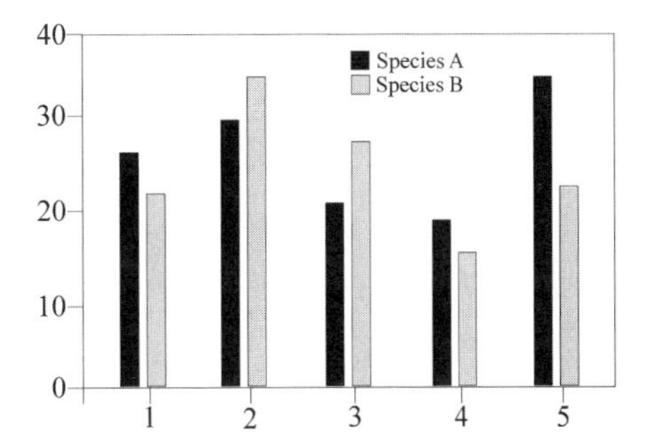

barley yellow dwarf virus (BYDV) virus; causative agent of the most widespread and economically important viral disease of cereal crops affecting more than 100 species and many wild grasses; transmitted via an aphid vector; controlled by the use of resistant plant lines and pesticides.

barophile a microorganism with optimal growth in environments with high hydrostatic pressure.

barotolerant a microorganism that can tolerate environments with high hydrostatic pressure but prefers conditions of 1 atmosphere.

barrier physical separation that protects the integrity of a product from the introduction of microbial contaminants.

barrier conditions physical, chemical, or biological obstacles that define populations of microorganisms in an environment; in contrast to boundary conditions.

basal salts solution/basal salts medium (BSM) a nutrient solution used in cell culture applications that contains the inorganic ions (e.g., sodium, calcium, magnesium, phosphate, sulfate) that are essential for the growth and maintenance of the cells.

base pair in DNA, the pair formed via a weak bond between either adenine and thymine or cytosine and guanine on opposite strands of the molecule.

basic stain reagent with a negatively charged chromophore that is attracted to positively charged cell material.

basidia plural term of basidium.

basidiomycete term descriptive of a grouping of ubiquitous fungi including puffballs, shelf fungi, mushrooms, rusts, and smuts; commonly isolated in soil and decaying plant material.

Basidiomycotina a subdivision of the Amastigomycota; grouping of saprophytic, symbiotic, or parasitic fungi that are characterized as unicellular or more typically with septate mycelium and producing basidiospores on the surface of a basidium.

basidiospore structure formed by approximately 1200 genera of fungi; generally not culturable on laboratory media although some will produce sterile mycelia; frequently observed in spore trap air samples especially during periods of high humidity or rain; some Type I allergies and Type III hypersensitivity reported with exposure.

basidium surface structure characteristic of members of the Basidiomycotina where basidiospores are produced.

basipetal developing toward the base; characteristic of a chain of conidia in which the youngest spore is located at the base; in contrast to acropetal.

BAT Branching Atmospheric Trajectory.

batch culture a fixed volume, closed system for the growth of organisms; in contrast to chemostat.

batch system model biological components and supportive nutrients are added to a closed system; a self-sustaining system when there is a suitable input of radiant energy and photoautotrophic organisms; nutrients are recycled within the microcosm.

bathypelagic zone designation for the vertical area in the marine environment that is generally aphotic and cold and extends below the epipelagic zone in the region from 200–6000 meters.

BAX an automated, commercially available polymerase chain reaction amplification method used for the screening of samples for the presence of *Listeria monocytogenes* and other pathogens.

Bchl *a* chlorophyll pigment used in photosynthesis by purple bacteria.

Bchl *b* chlorophyll pigment used in photosynthesis by purple bacteria.

Bchl *c* chlorophyll pigment used in photosynthesis by green bacteria.

Bchl *d* chlorophyll pigment used in photosynthesis by green bacteria.

Beauveria fungal genus; cottony to wooly colony on malt extract agar that is white in color becoming pale yellow or pink on the surface and pale on the reverse; the hyaline hyphae are septate with dense masses of obclavate conidiogenous cells that terminate in a thin, zigzagging filament; conidia are small, hyaline, and round or oval shaped cells that are often not observed with spore trap sampling; some species are insect pathogens.

beef extract a chemically undefined proteinaceous material used to desorb viruses from membrane filters; at acidic pH, the organic material precipitates out of solution, and after centrifugation, viruses can be recovered from the pellet.

Beggiatoa bacterial genus; chemolithotrophic members of the Cytophagales that form filaments, oxidize H_2S, and deposit sulfur intracellularly when growing on hydrogen sulfide; colorless sulfur bacteria isolated from outflow of sulfur springs.

BEI biological exposure index.

Beijerinck, Martinis (1851–1931) Dutch microbiologist who recognized the ubiquity of microorganisms and the selective influence of the environment, isolated symbiotic and non-symbiotic aerobic nitrogen fixing organisms and sulfate reducing organisms, and with Winogradsky developed the enrichment culture method in 1901.

Beijerinckia bacterial genus; member of the Azotobactereaceae; nitrogen fixing free-living, pear-shaped bacilli with large lipid bodies located at each end; produces an extensive slime; four recognized species; isolated from acidic soils.

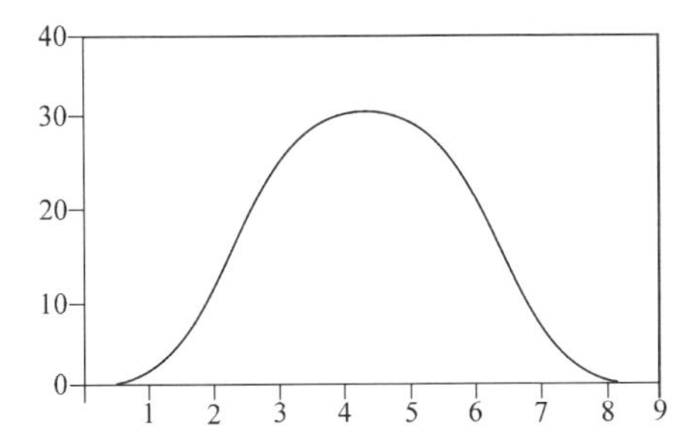

bell curve plot of data with a Gaussian distribution.

Beltsville process a method used in composting in which air is suctioned through perforated pipes buried inside the pile resulting in partial oxygenation of the material; in contrast to the Rutgers process.

benthic/benthos the bottom of the ocean or a lake representing the interface of the hydrosphere and the lithosphere.

Bergey's Manual of Systematic Bacteriology a multivolume publication listing the taxonomic classification of the bacteria and Archaea based on classical and molecular information and organized by dichotomous keys.

beta error Type II error.

betaine a colorless, water-soluble compound ($C_5H_{11}NO_2$) that is used as an osmolyte.

beta oxidation the oxidation of fatty acids by the splitting off of two carbons at the same time; the enzymes used by prokaryotes are in the cytoplasm while in eukaryotes they are located in the mitochondria.

BGMK cells buffalo green monkey kidney cells.

B horizon the illuvial horizon of the soil where deposition has occurred and there is maximal accumulation of iron oxides, aluminum oxides, and silicate clays; region beneath the A horizon and above the C horizon.

bhp designation for the polychlorinated biphenyl degradation genes present in gram-negative and gram-positive bacteria that encode for enzymes used in the conversion of biphenyl and polychlorinated biphenyl into benzoate and chlorobenzoates.

bias a systematic error in the design, conduct, or analysis of a scientific study that results in misinformation generally classified as selection bias, surveillance bias, or misclassification bias.

biciliate having two cilia.

bidirectional acetogen acetogenic bacterium capable of growth using the acetyl-CoA pathway in both directions resulting in either acetate or H_2 as the reduced end product.

Bifidobacterium bacterial genus; gram-positive, obligate anaerobe that does not produce filaments, but coryneform cells are common; normal flora in the gastrointestinal tract of humans; several species act as probiotics producing oligosaccharides that can be used as dietary supplements to modulate colonic microflora.

Bifidobacterium lactus bacteria species; increased acid tolerance and increased oxygen levels; used in infant formula and yogurt.

biflagellate having two flagella.

bifurcate forking with two branches.

bimodal two peaks; representation of data demonstrating that two categories have the same frequency.

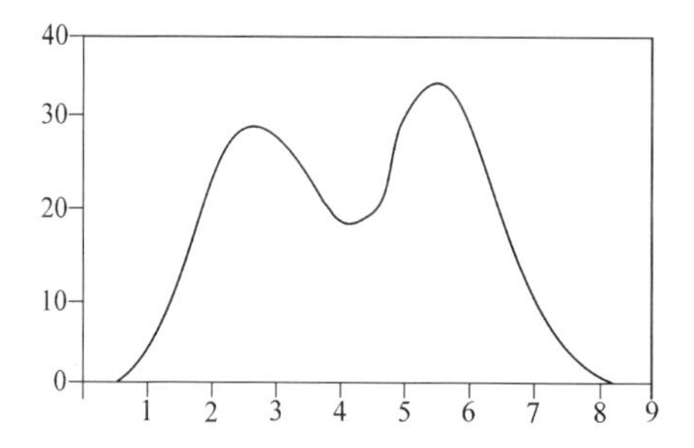

binomial distribution the probability of different outcomes for a series of random events in which only one of two values is possible; this distribution resembles a Gaussian distribution as the number of samples increases.

binomial system of nomenclature taxonomic listing of microorganisms using a genus and a species name.

binucleate having two nuclei.

bioaccumulation the accumulation or increase in concentration of a substance in an organism over time.

bioaerosol a suspension of biological particles or microbial metabolites in the air; individual particles or droplets are generally 0.5–20 μm in diameter and transported through the air from liquids or surfaces; may result in exposure via inhalation.

bioassay a biological-based analysis.

biochemical oxygen demand (BOD) an estimate of the readily useable concentrations of organic matter in a medium; determined by allowing microorganisms to consume oxygen in a solution of organic matter in a sample; a high value indicates an abundance of organic matter.

biocidal an agent capable of killing bacteria and other organisms.

biocontrol agent microorganism used to suppress the activity of other organisms; generally used to describe microorganisms applied to seed or planting material to combat soil-borne phytopathogens.

bioconversion the use of microorganisms in an industrial setting to biochemically convert a substance into another substance.

biodegradation the breakdown of a substance into smaller products as a result of microbial activity; generally viewed as a desirable process in the cycling of materials, including nutrients and contaminant compounds in the environment; in contrast to biodeterioration.

biodeterioration the undesirable transformation of a substance as a result of biological activity; minimizing this process can be achieved by destroying or removing microorganisms responsible for the activity, keeping the substance in an environmental condition that minimizes microbial activity, or modifying the substance with additives to reduce its availability as a substrate for microbial activity.

biodisk a rotating circular platform with a film of microorganisms used in some secondary wastewater treatment facilities.

biodiversity the presence of a wide range of biological species in a defined area or at a specific time.

bioemulsan/bioemulsifier biologically derived surfactant used in industrial and commercial activities for the enhanced recovery of oil, bioremediation of oil-polluted water and soil, formulation of pesticides and herbicides, production of oil/water emulsions in foods, and in cosmetics; low molecular weight agents are generally glycolipids while high molecular agents are amphipathic polysaccharides, lipopolysaccharides, proteins, and lipoproteins.

biofilm a community of microorganisms (e.g., algae, bacteria, fungi, and protozoa) that develops on surfaces in aqueous environment; the microorganisms are anchored to the surface and held together by a polysaccharide material produced and secreted by some of the microorganisms; the activities of these microorganisms may be beneficial (e.g., biodegradation) or detrimental (e.g., clogging of drinking water distribution pipes).

birefringent demonstration of a different refractive index based on the plane of polarization of light.

biogenic produced by living organisms.

biogeochemical involving biological, geological, and chemical aspects.

biogeochemical cycling the conversion of materials in the ecosphere through the characteristic pathways of the carbon cycle, the hydrogen cycle, the oxygen cycle, the sulfur cycle, the phosphorus cycle, and the iron cycle resulting in physical and chemical transformations that are mediated by biochemical activity.

biogeochemistry the study of chemical transformations of elements of geochemical interest that are mediated by microbial activity, such as nitrogen cycling and sulfur cycling.

biohazard biological material that is likely to cause a risk to human health.

biohazard bag thick plastic autoclave bag, generally red in color, that is stamped with the international biohazard symbol; used for the disposal of biological materials.

biohazard symbol

bioinformatics the use of computer database systems in the analysis of DNA and protein sequences.

bioleaching the solubilization of compounds from a solid material into the surrounding liquid as a result of microbial activity.

Biolog a computer-based identification system that utilizes a 96-well microtiter plate for substrate utilization assay and colorimetric end-point analysis to identify bacteria and yeast.

biological acid-producing potential test (BAPP test) an assay designed to determine if bacteria can maintain acidic conditions and be self-sustaining through the generation of sulfuric acid by the oxidation of sulfides in a mineral sample.

biological control the use of an organism to affect a change in the presence of another organism; commonly used for insect control in agriculture.

biological exposure index (BEI) procedures of the American Conference of Governmental Industrial Hygienists for estimating the amount of a substance in a human body by measuring for it in tissue, body fluids, or exhaled air.

biological half-life time required for a living organism to eliminate one-half of a substance from its body.

biological magnification a phenomenon in living systems whereby the concentration of persistent, lipophilic chemical compounds is increased over time because they accumulate in organisms that are ingested by other organisms higher in the food chain.

biological oxygen demand biochemical oxygen demand.

biological plausibility the likelihood that biological processes are involved in an event although no scientific proof is yet available; the coherence of a theory with the current body of scientific knowledge.

biological safety cabinet laminar flow chamber with HEPA filtration that is designed to minimize the exposure of a user to biological materials present in cabinet.

biological vector a living object that serves to transfer an infectious agent to a new host, while in association with the vector the infectious agent replicates sufficiently to increase its concentration.

biological warfare agent/biowarfare agent microorganism or microbial by-product with the potential for use as a weapon of mass destruction; generally agents are easy to propagate with minimal instrumentation and are easy to disseminate via an inhalation, ingestion, or dermal route of exposure; a low infectious dose or low threshold limit value for a broad range of individuals would increase the impact on the exposed population.

biologic marker biomarker.

bioluminescence the generation of visible light by microorganisms.

biomagnification biological magnification.

biomarker the presence of a biological entity that signals the presence or occurrence of a specific event or exposure.

biomass the amount of living material present; a measurement of the quantity of energy being stored in a segment of the biological community expressed in units of weight.

biometer flask a commercially available Erlenmeyer flask-shaped device with a side arm that contains fluid for trapping of CO_2 that is used in batch experiments involving the monitoring of the disappearance of a single chemical over time.

biomineralization the transformation of a substance into a mineral that is mediated by a biological process.

biomonitor indicator organism.

biopesticide microbial pest control agent.

biopreservation the use of microorganisms or microbial by-products to prevent spoilage and extend the shelf-life of foods.

biopsy removal of tissue or cells from a living organism for examination.

bioreactor a vessel in which biological processes are used to convert a substrate into desired end products.

bioremediation the use of microorganisms to transform a chemical contaminant into a less harmful or harmless chemical.

biosolid the solid fraction of the material produced during treatment of domestic wastewater.

biosafety levels the classification of laboratory environments based on the risk of human disease with exposure to known microorganisms; currently four levels are recognized (BSL1, BSL2, BSL3, and BSL4).

biosecurity concern for the biological contamination of food, air, and water, and the processes and procedures implemented to minimize or prevent such contamination.

biosensor a microorganism that is used to signal the presence of a specific compound or condition because it displays an easily measurable response upon exposure to that compound or condition.

biostatic inhibits the growth of bacteria.

biosynthesis the formation of cellular components from simpler molecules; also used to describe the synthesis of metabolic by-products through biotechnology applications.

biotechnology the use of microorganisms or microbial reactions in a defined chemical process to produce a desirable product.

bioterrorism the use of a biological agent to promote a political or criminal agenda.

Biotest sampler a portable, battery-powered impactor sampler for the collection of bioaerosols for follow-on culture analysis that uses agar-filled wells positioned on a strip rather than an agar-filled plate.

biotic involving living organisms.

biotin an essential growth factor for many cells; widely used as a label for macromolecules which can then be detected by high affinity binding of labeled avidin or streptavidin.

biotype a variation of a microbial species or serotype.

biphasic having two phases.

biphasic growth curve the growth curve of a microbial population in culture during diauxic growth.

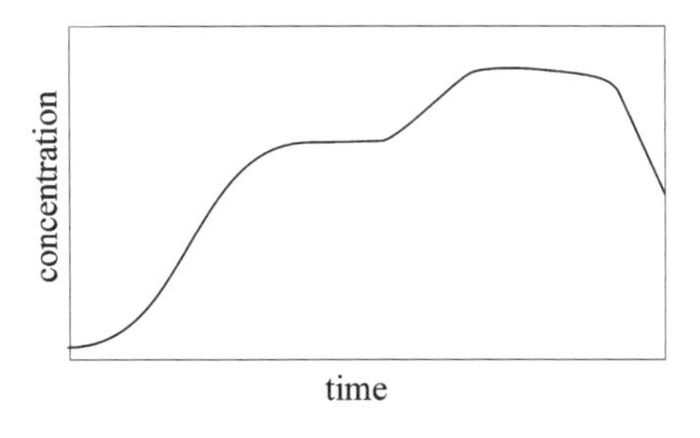

biplate segmented dish designed for the use of two different culture media.

Bipolaris fungal genus; downy colony on malt extract agar that is whitish in color becoming dark olive to black on the surface and reverse; brown hyphae are septate with brown geniculate conidiophores producing brown, fusoid, distoseptate (3–6 cells) poroconidia that is characterized by a scarcely protuberant hilum and germ tubes that develop from the terminal cell of the conidium; in contrast to *Curvularia*, *Dreshlera* and *Exserohilum*.

biseriate the arrangement of phialides supported by metulae that are characteristic in the speciation of the genus *Aspergillus*.

black aspergilli general grouping of *Aspergillus* species that bear dark or black spores; representative organisms produce extracellular enzymes and metabolites useful in the fermentation industry.

black death nonscientific terminology descriptive of the human suffering caused by *Yersinia pestis*; the plague.

black mold nonscientific terminology descriptive of dark colored fungi; often used to describe fungi that are present on water damaged building materials.

black smoker hot vent.

blank a representative of the sample matrix without the analyte present.

blastic descriptive of the budding formation of some fungal conidia.

blastoconidium a spore produced by budding; in yeast this term is used to describe a bud.

Blasto D medium fungal culture medium used to convert *Blastomyces dermatitidis* from the filamentous phase to the yeast phase.

Blastomyces fungal genus; a pathogenic thermophilic diphasic fungus with a yeast form at 37° and a mycelial form at 23°C.

Blastomyces dermatitidis fungal species; a diphasic fungus with a white mycelial stage growing on Sabouraud dextrose agar at 25–30°C although some isolates may produce a dark pigment in the growth medium; older colonies become tan to brown; smooth-walled, spherical or oval conidia (2–10 μm) are found on terminal ends of hyphal branches; the yeast form (8–15 μm) is present as thick, refractile broad-based budding cells cultured at 37°C; causative agent of blastomycosis in mammals.

blastomycosis disease caused by inhalation of spores of *Blastomyces dermatitidis*.

blight plant disease in which a general withering occurs without rotting of the tissue.

blind/blinding processing and analyzing samples without knowledge of the identification of individual samples as test or control material.

BLIS bacteriocin-like inhibitory substances.

blocking agent a substance used to mask binding sites and prevent attachment.

blood-glucose-cysteine agar a fungal culture medium used to promote conversion of diphasic fungi such as *Histoplasma capsulatum*, *Blastomyces dermatitidis*, *Paracoccidioides brasiliensis*, and *Sporothrix schenckii* to the yeast phase.

bloom rapid, visible growth of microorganisms, generally used to describe the growth of algae and cyanobacteria in surface water, generally as a result of a dramatic change in nutrient availability or environmental conditions; may result in production of harmful levels of toxins, taste and odor-producing compounds (e.g., geosmins), and eutrophication.

blotting transfer of material from a gel matrix to a membrane.

blue-green algae cyanobacteria.

blue-green bacteria cyanobacteria.

blue oxidases multi-copper enzymes including laccases, ascorbate oxidases, and vertebrate ceruloplasmin that are produced by many plants and fungi; may have industrial applications for detoxification of phenolic compounds and azo dyes.

BOD biochemical oxygen demand.

body burden total amount of a chemical that is retained within a body.

boiling point temperature at which a liquid changes into a vapor.

BoNT botulinum neurotoxin.

Borrelia bacterial genus; member of the Spirochaetales that are not very strict anaerobes; transmitted via the bite of ticks and lice; causative agents of relapsing fevers in humans.

Borrelia burgdorferi bacterial species; causative agent of Lyme disease.

Borrelia recurrentis bacterial species; causative agent of louse borne recurrent fever.

botanical epidemiology the study of the spread of plant diseases within a crop and field-to-field within a region.

botryose clustered, as grapes.

Botrytis fungal genus, broadly spreading light gray to brown colony on malt extract agar with generally solitary, brown, erect conidiophores often with black sclerotia; pale brown, globose, ovate, or ellipsoidal smooth-walled conidia; currently 25 recognized species; isolated from soil, stored and transported fruit and vegetables, and indoors with house plants; may be saprophytic, but pathogenic to higher plants, flowers, leaves, stems, fruit, and associated with blight of grapes, strawberries, lettuce, cabbage, and onions; water activity of 0.93–0.95.

Botrytis cinerea fungal species; formation of conidiophores in irregular patches with obovoid conidia having a protuberant hilum; worldwide, but found mainly in humid temperate and subtropical areas; disease-causing organism that is generally associated with grapevines, strawberries, and other fruits and produce, but it does not affect healthy leaves; termed "noble rot" as grapes infected with this organism impart pleasing flavors to wine.

bottom yeast a brewery yeast, generally *Saccharomyces carlsbergensis*, that settles to the bottom of the fermentation liquid during processing, in contrast to a top-fermenting yeast.

botulism a disease resulting from exposure to the toxin produced by *Clostridium botulinum*; exposure results from ingestion of contaminated food or infection of an exposure wound.

boundary conditions the physical and biochemical limitations of microorganisms that define their establishment in an environment; in contrast to barrier conditions; in mathematical modeling this refers to the limitations imposed on a system by the environment in which the system exists.

boundary layer the narrow area at the interface between a solid surface and a liquid or air flowing over a surface; the velocity of the fluid or air may be greatly reduced in this area.

bovine serum albumin (BSA) albumin from the serum of cattle; used as a source of nutrients for the growth and maintenance of cells *in vitro*, such as in cell culture; also used as a blocking agent in polymerase chain reaction amplification methods and as standards for protein assays.

box plot/box and whisker plot graphical representation of data in which the median is designated with a + and a box is drawn around the upper and lower quartiles of the data so that the middle 50% of cases are arranged within the range of scores defined by the box; the variability of the data determines the box length and the placement of the + demonstrates if the data are skewed; the whiskers are lines drawn from the box indicating the highest and lowest data points one step away from the quartiles.

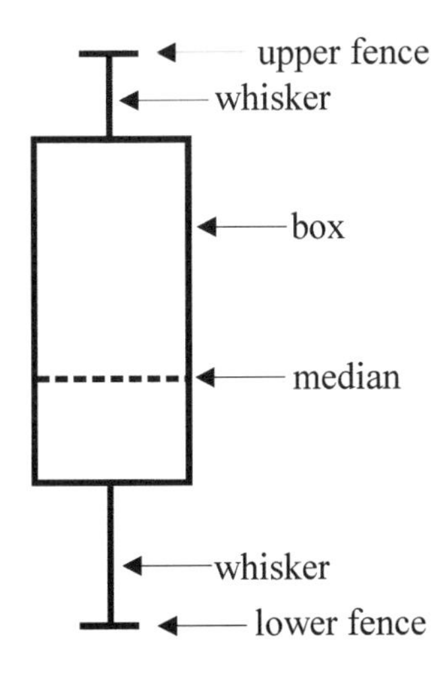

Bradyrhizobium bacterial genus; gram-negative, motile, nitrogen-fixing bacteria that live in a symbiotic relationship in nodules on the roots of leguminous plants, such as soybeans, peas, and clover.

Branching Atmospheric Trajectory model computer-based program used to calculate the trajectory of airborne fungal spores of agricultural importance.

breakpoint chlorination the input of chlorine into drinking water, swimming pools, or sewage effluent for disinfection at an approximate molar ratio of two moles of chlorine to one mole of ammonia, present in the effluent as ammonium nitrogen, resulting in the conversion of the ammonia to molecular nitrogen and a lowering of the biological oxygen demand of the effluent.

breathing zone conventional distance from the floor (1.5 m) used as the area of human exposure from the respiratory route of transmission.

brewing the production of alcoholic beverages by microbial fermentation.

BRI building related illness.

bright-field microscopy the most commonly used technique for microscopic examination of microbiological specimens; technique uses contrast between images as the basis for visualization and specimens are often stained (e.g., Gram reaction/Gram stain) to enhance the appearance of objects.

broad spectrum antibiotic an antibacterial compound that is effective against both gram-positive and gram-negative organisms.

Broad Street pump location of the pump responsible for the distribution of bacterial contaminated drinking water that resulted in the spread of cholera in London; first documented waterborne outbreak of disease caused by microorganisms.

bromuron 3-(*p*-bromophenyl)-1-methoxy-1-methylurea, a herbicide that undergoes acetylation by soil fungi.

bronchitis inflammation of the mucous membranes of the bronchi.

Brownian motion/Brownian movement the random motion of particles resulting from intermolecular collisions.

brown rot fungi category of basidiomycetes that attack the cellulose in wood, but either do not attack the lignin or only partially degrade it; in contrast to the white rot fungi.

brush border membrane vesicles (BBMVs) used in Bt binding assays.

bryostatins a family of macrocyclic lactones with anticancer activity.

BSA bovine serum albumin.

BSL1 a biosafety level designated as a laboratory environment that is suitable for work involving well-characterized organisms that are not known to cause disease in healthy adult humans, and those that have minimal potential hazard to laboratory personnel and the environment; a laboratory at this level does not have to be separated from the general traffic patterns in the building and the work performed is generally conducted on open bench tops using standard microbiological practices; no special containment equipment or facility design is required or generally used; laboratory personnel generally wear a lab coat, gloves and protective eyewear; personnel have specific training in the procedures conducted in the laboratory and are supervised by a scientist with general training in microbiology or a related science.

BSL2 a biosafety level that is similar to BSL1 and is suitable for work involving agents of moderate potential hazard to personnel and the environment, but laboratory personnel have specific training in handling pathogenic organisms and the access to the laboratory is limited when work is being conducted; in addition, extreme precautions are taken with contaminated sharp items, and procedures in which infectious aerosols or splashes may be created are conducted in biological safety cabinets or other physical containment equipment; laboratory personnel should wear a lab coat and two pairs of gloves are recommended when handling hazardous materials.

BSL3 a biosafety level associated with clinical, diagnostic, teaching, research, or production facilities where indigenous or exotic microorganisms that may cause serious or potentially lethal disease due to an inhalation route of exposure are handled; laboratory personnel have specific training in handling pathogenic and potentially lethal organisms, and are supervised by scientists with experience in working with these organisms; all procedures involving the manipulation of infectious materials are conducted within biological safety cabinets or other physical containment devices or by personnel wearing appropriate personal protective clothing and equipment; the laboratory has special engineering and design features with restricted access; laboratory personnel wear gloves, coveralls or gown, and facemask with respiratory filter.

BSL4 the biosafety level that is required for work with dangerous and exotic microorganisms that pose a high individual risk of infections transmitted via aerosol that result in life-threatening disease; in addition, organisms that have a close or identical antigenic relationship to BSL4 organisms are also handled at this level until sufficient data are obtained either to confirm continued work at this level, or to work with them at a lower level; laboratory personnel have specific and thorough training in handling extremely hazardous infectious agents and they understand the primary and secondary containment functions of the standard and special practices, the containment equipment, and the laboratory design characteristics; access to the laboratory is strictly controlled by the laboratory director and the facility is either in a separate building or in a controlled area within a building, which is completely isolated from all other areas of the building; the laboratory has special engineering and design features to prevent microorganisms from being disseminated into the environment and all activities are confined to Class III biological safety cabinets, or Class II biological safety cabinets used with one-piece positive pressure personnel suits ventilated by a life support system.

BSM basal salts medium.

Bt Bacillus thuringiensis.

BTEX acronym for a group of aromatic compounds (benzene, toluene, ethylbenzene, and xylene).

Bt transgenic crop agricultural plant with *Bacillus thuringiensis* genes inserted in an effort to confer resistance to a specific pest.

budding a means of asexual reproduction in which an outgrowth of the parent cell develops to form a daughter cell; commonly observed in yeasts.

budding bacteria member of this group of bacteria grow in low nutrient environments with the assistance of prosthecae which afford the cell greater efficiency in concentrating available nutrients.

buffalo green monkey kidney cells (BGMK) cells from the kidneys of green monkeys established in Buffalo, New York that have been adapted to grow *in vitro* as a continuous cell line; used for the detection and cultivation of enteroviruses.

buffer solution that serves to minimize changes in pH.

buffered yeast charcoal extract (BYCE) an agar medium used for the isolation of *Legionella*; addition of bromocresol blue or purple as indicator dyes may be used to indicate certain species; some species require bovine serum albumin amendment; antibacterial and antifungal amendments may be added to minimize growth of other organisms.

building related illness (BRI) a condition when medically recognized adverse health effects are caused by exposure indoors to a specific causative agent; in contrast to sick building syndrome.

bulk sampling collection of material directly from a matrix; in contrast to a swab sampling.

Buller's drop a suspension of fluid that forms on the hilar appendage of a sterigma resulting from fungal secretions and moisture from the atmosphere that assists in the active fungal spore dispersal of basidiospores.

bunsen burner laboratory bench top gas burner used to direct a narrow flame for flame sterilization.

buoyant density centrifugation technique used to separate constituents in a solution based on their density.

buried slide technique placement of a glass microscope slide in soil or sediment to obtain a sample of microbial populations present.

Burkard personal impactor sampler a battery powered impactor sampler designed for the collection of airborne fungal spores and pollen; particles are collected through a slit at 10 liters/min. onto a sticky surface and analyzed using light microscopy.

Burkard single-stage impactor sampler a compact, battery powered impactor sampler designed for the collection of bioaerosols at a flow rate of 28.3 liters/min. through precision drilled holes onto an agar-filled petri plate for follow-on culture analysis.

Burkholderia bacterial genus; aerobic, chemoorganotrophic bacilli with polar flagella; some species are pathogenic to humans and some are phytopathogens.

Burkholderia cepacia bacterial species; potential opportunistic pathogen with environmental exposure via aerosols generated during wastewater treatment practices; phytopathogen, causative agent of onion bulb rot; formerly termed *Pseudomonas cepacia*.

***Burkholderia* LB-400** mesophilic bacterial strain that can degrade polychlorinated biphenyls (PCBs).

burst size the number of viruses released when an infected cell is lysed.

bursiform bag-shaped or pouch-like.

BYCE buffered yeast charcoal extract.

BYDV barley yellow dwarf virus.

byssinosis debilitating respiratory condition characterized by chest tightness resulting from the inhalation exposure of dust, fiber, or endotoxin.

C

cable tool drilling a drilling technique used to obtain samples in shallow and deep sediment environments commonly used for the drilling of water wells.

Caco-2 cells epithelial cells derived the human intestine that have been adapted to grow *in vitro* as a continuous cell line; used for the detection and cultivation of microorganisms such as hepatitis A virus and *Cryptosporidium parvum*.

CAEPR Community Action Emergency Planning Response.

caespitose in dense groups forming tufts.

calcium cycle the transformation of calcium from one form to another in reservoirs by biological activity.

calibrate/calibration the adjustment of an instrument by comparison to standard reference material.

calibration standards materials to be used for verifying the performance of an instrument.

calicivirus viral genus; in the family Caliciviridae containing many species, including the Norwalk virus, which is responsible for many water- and foodborne outbreaks of gastroenteritis.

Calvin cycle autotrophic CO_2 fixation.

calyptrate having a cap or lid.

campanulate bell-shaped.

Campylobacter bacterial genus; gram-negative, microaerophilic, *chemoorganotrophic*, motile with a corkscrew motion, spiral-shaped curved bacteria; many species are pathogenic to humans and animals; isolated from the intestinal tract, oral cavity and reproductive organs of humans and animals.

Campylobacter jejuni bacterial species; ubiquitous in the environment; causative agent of acute bacterial enteritis resulting from an ingestion route of exposure, often due to the consumption of contaminated chicken, but also associated with contaminated potable water sources.

canaliculate having grooves or channels.

candle jar low cost, older method used to provide a reduced oxygen atmosphere for incubation of cultures using a glass screw-capped vessel with a small candle that is naturally extinguished when the oxygen level is decreased to very low levels.

canker a localized necrosis on a plant resulting in a lesion, usually observed on the stem; symptom of some plant diseases caused by microorganisms.

capneic increased CO_2.

capsid the protein coat of a virus that encloses the nucleic acid; generally either icosahedral or helical in shape.

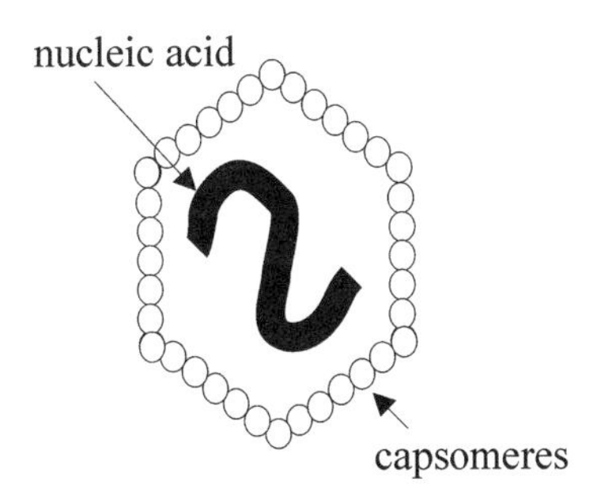

capsomere a protein subunit in a viral capsid; the cluster of viral proteins or protomers.

capsule a glycocalyx structure around a cell that is rigid and excludes particles.

carbaryl 1-naphthyl-*N*-methylcarbamate, a pesticide that undergoes biodegradation readily by microorganisms in the soil; in contrast to aldrin.

carbol fuchsin an aniline dye used to stain *Mycobacterium* cells red for viewing with light microscopy.

carbonaceous relating to, containing, or composed of carbon; having a black and brittle appearance.

carbonaceous biological oxygen demand (CBOD) amount of oxygen consumed by microorganisms for the biochemical oxidation of organic material.

carbon cycle the transformation of carbon in reservoirs of atmospheric CO_2, inorganic carbon dissolved in seawater, organic carbon in biomass in terrestrial and aquatic environments, and fossil fuels and sedimentary rock through a series of biogeochemical processes; fixation of CO_2 to form organic compounds is conducted by autotrophic activity of the primary producers, mostly by photosynthesis of the algae, green bacteria, and purple bacteria, while some activity is the result of chemolithotrophic microorganisms; a portion of the organic carbon is converted back to CO_2 during respiration of the primary producers and the remainder is the net community productivity which is transferred from one trophic level to another by grazing and predation through a variety of food chains; fermentation and methanogenesis occur during anaerobic conditions.

carboxydotrophic bacteria bacteria that aerobically utilize carbon monoxide as both a carbon source and as an energy source, for example, *Pseudomonas carboxidoflava* and *Pseudomonas carboxidohydrogena*.

carboxysomes enzymatic polyhedral-shaped inclusion bodies necessary for the Calvin cycle; first purified and studied in *Halothiobacillus neapolitanus*, formerly termed *Thiobacillus neapolitanus*.

carcinogen/carcinogenic an agent that causes the initiation of a tumor; aflatoxins produced by some fungi have been implicated as carcinogenic agents.

carcinoma a malignant tumor that arises in epithelial cells, infiltrates the surrounding tissue, and spreads to distant sites in the body.

cardinal temperature the three temperatures (i.e., minimum, optimum, and maximum) for growth that are characteristic of a particular microorganism; these temperatures are not completely fixed but are influenced by the composition of the growth substrate.

cardiovascular system (CVS) heart and blood vessels.

carnitine an amino acid ($C_7H_{15}NO_3$) that is used as an osmolyte.

carotenoid pigment coloration, generally yellow, orange, red or purple, found in some bacteria and algae, that affords some protection from ultraviolet radiation.

carrier an individual that transmits an infectious agent to others, but does not show any symptoms of the disease.

carrying capacity the maximum amount of a substance that can be present without resulting in an adverse effect.

CAS number Chemical Abstracts Service number.

case control study investigation in which attributes of interest are compared between two groups, one group of individuals having been identified with a disease or adverse health effect and a similar group identified as a control, without the illness.

case-fatality rate the number of individuals who die of a disease after being diagnosed divided by the total number of individuals with the disease; in contrast to mortality rate.

case study a survey in which no control group is studied for comparison; in contrast to case control study.

catabolic/catabolism the breakdown of organic or inorganic compounds resulting in the release of energy.

catabolite repression the suppression of enzymatic reactions when an organism is grown in a glucose-containing medium.

catalyst a substance that increases the rate of a chemical reaction by decreasing the activation energy needed for the reaction to occur without the substance itself being used up.

catenulate arranged in chains.

Caulobacter bacterial genus; characteristic spiral shape that use a flexing motion for motility; the appendages of these budding bacteria are referred to as stalks.

causal relationship/causation the association of a factor with an outcome.

causative agent/causal agent that which elicits the adverse health effect resulting in disease.

CBOD carbonaceous biological oxygen demand.

CCD camera charge-coupled device camera.

CCL contaminant candidate list.

CDC Centers for Disease Control and Prevention.

CDFF constant-depth film fermenter.

cDNA complementary DNA.

CEGLs continuous exposure guidance levels.

ceiling maximum allowable exposure concentration for an airborne substance that is not to be exceeded.

cell the fundamental unit of life; structural subunit of living organisms that is separated from the environment by a delimiting membrane.

cell culture maintenance and growth of cells in an *in vitro* culture using appropriate nutrients; cells have finite division potential; used for the propagation and detection of viruses and parasites.

cell line a permanently established cell culture that can proliferate indefinitely if provided with the appropriate nutrients and environmental conditions.

cell lysis the disintegration of cellular structure and function.

cell-mediated immune reaction an immune response generated by the activity of non-antibody-producing cells; descriptive of the immune response resulting in hypersensitivity pneumonitis caused by the inhalation route of exposure to airborne thermophilic actinomycetes.

cellulase enzyme system a series consisting of a C_1 enzyme, a C_x enzyme, and a β-glucosidase enzyme that degrade various forms of cellulose.

cellulolytic utilizes cellulose as a carbon source for metabolism.

cellulose a carbohydrate consisting of β(1-4)-linked glucose units in a linear chain that can be degraded aerobically by a variety of fungi resulting in the formation of CO_2, water and cell biomass, and anaerobically primarily by *Clostridium* spp. resulting in low molecular weight fatty acids, CO_2, water and cell biomass.

cell wall structural material that defines the shape of a cell; present in prokaryotes and many eukaryotes, including plants and fungi, but not in animals.

CEMS continuous emission monitoring systems.

central nervous system (CNS) brain and spinal cord.

Centers for Disease Control and Prevention (CDC) the federal agency of the United States within the Department of Health and Human Services with the mission to promote health and quality of life by preventing and controlling disease, injury, and disability and to protect the health and safety of people by providing information to enhance health decisions.

centrifuge instrument that utilizes centrifugal force generated by high speed spinning of containers arranged around an axis for the sedimentation and separation of materials in liquid within those containers.

C_1 enzyme an enzyme in the cellulase enzyme system that is active on native cellulose but not partially degraded cellulose molecules.

C_x enzyme an enzyme in the cellulase enzyme system that cleaves partially degraded cellulose polymers either by internally breaking the chain at random resulting in the formation of cellobiose and oligomers using endo–β-1, 4-glucanases or by using exo-β-1,4-glucanases to attack the polymer at the end of the chain to form cellobiose.

Cephalosporium acremonium former terminology for *Acremonium*.

cerebriform having brain-like folds.

cesium chloride (CsCl) a salt that produces high-density aqueous solutions; used in cesium chloride density-gradient centrifugation to separate microorganisms, particles, and DNA molecules of different densities.

cesium chloride density-gradient centrifugation a method for the determination of density of a molecule of DNA that is based on the nucleotide content and conformation (linear or circular). DNA molecules of different densities can be separated in the same configuration.

CFM cubic feet per minute.

CFSC continuous-flow slide culture.

CFU colony forming unit.

CFU/m^3 measurement used in aerobiology as the number of colony-forming units of bacteria or fungi cultured per cubic meter of air sampled.

Chaetomium fungal genus; grows rapidly but may require 1-3 weeks for sporulation on laboratory media; colony on malt extract agar is wooly in texture with white color that turns gray to olive on the surface with a pale yellow to brown reverse; septate hyphae pale brown in color; unicellular ascospores are formed within a round to oval, brown to black perithecia with distinctive long brown setae that mask the ostiole; members of this genus are cellulolytic and readily isolated from soil, seeds, and cellulosic substrates especially wet wallboard paper, dung, wood and straw materials; used in the production of cellulase for industry.

Chaetomium globosum fungal species; distinctive small, brown lemon or football-shaped ascospores; production of chaetoglobosin mycotoxins and a variety of mutagens.

chain-of-custody tracking procedures to verify the handling of samples from collection through reporting of the data.

chaotropic a substance that disrupts the structure of water, thereby promoting the solubility of nonpolar substances in a polar solvent (such as urea, Tween 80, or ethanol in water); by disrupting the structure of water and keeping viruses in solution, these compounds decrease the adsorption of some viruses to surfaces; in contrast to antichaotropic.

Chelex a resin used in sample pretreatment to bind metal ions that inhibits the polymerase chain reaction amplification reaction.

Chemical Abstracts Service number (CAS number) a unique means of identification in which each chemical is denoted by a number.

chemical cabinet storage apparatus for laboratory chemicals.

chemical fume hood a laboratory apparatus designed to exhaust volatile compounds through appropriate filters before releasing the air to the outdoors.

chemical oxygen demand (COD) index of the amount of total organic carbon determined by measuring the amount of oxidizing reagent consumed during oxidation of organic matter using dichromate or permanganate; in contrast to biochemical oxygen demand.

chemiluminescence the release of light during a chemical reaction.

chemiluminescent *in situ* hybridization (CISH) rapid detection and identification chemiluminescence method used with peptide nucleic acid probes; can provide results in the analysis of bacterial contaminants in water samples within one day.

chemiosmosis the use of a proton gradient across cellular membranes to couple enzymatic reactions and to generate ATP.

chemocline the vertical layer in an aqueous solution in which the chemical conditions change rapidly and significantly compared to the layers above and below.

chemoheterotroph/chemoheterotrophic a microorganism that utilizes organic compounds as the carbon source.

chemolithoautotroph/chemolithoautotrophic a microorganism that is able to utilize CO_2 as the sole carbon source and utilizes an inorganic substrate as the energy source.

chemolithotroph/chemolithotrophic a microorganism that obtains energy through the oxidation of inorganic compounds (e.g., *Beggiatoa*); in contrast to a chemoorganotroph.

chemoorganotroph/chemoorganotrophic a microorganism that obtains energy through the oxidation of organic compounds; in contrast to a chemolithotroph.

chemostat a continuous culture device with a controlled dilution rate and addition of nutrients, also called a flow-through system; in contrast to batch culture.

chemosterilant a chemical used for sterilization; generally used for medical equipment and plastic labware, but rarely used for food processing due to the toxicity and taste.

chemotaxis movement of a microorganism in response to a chemical gradient.

chemotherapeutic agent a bactericidal or bacteriostatic agent that is used to selectively limit bacterial growth within a host.

chemotherapy treatment of an illness using chemicals or antibiotics.

chemotroph/chemotrophic an organism that utilizes a chemical as an energy source.

chi-squared (χ^2) test statistical analysis used to determine if data are different from expected values.

chitin a β(1-4)-linked polymer of *N*-acetyl-D-glucosamine that is used to estimate fungal biomass as it is present in the cell wall of many fungi but not found in plants or other soil microorganisms, although it is found in the exoskeleton of arthropods.

chlamydias gram-negative bacteria that are obligate intracellular parasites causing respiratory and urinary-genital tract diseases in humans.

chlamydospore a vegetative, asexual fungal spore often having an inflated, rounded, thick walled morphology and located within or at the terminal end of hyphae.

chloramine the reactive compound of chlorine and ammonia that is used for water disinfection that does not produce as many disinfection by-products as chlorine; generally is not as effective as chlorine for water disinfection except in drinking water distribution systems.

chloramphenicol a bactericidal agent that can be used as an amendment to fungal or algal culture media to suppress the growth of bacteria.

chlorination the addition of chlorine or chlorine-based compounds to disinfect drinking water and wastewater.

chlorine a strong oxidizer used as a chemical disinfectant for water.

Chlorobium bacterial genus; characterized as moderately thermophilic green sulfur bacteria.

chloroform a chemical used in conjunction with phenol to extract DNA or RNA from cells or used with methanol to extract lipids from cells.

chloroplast a chlorophyll-containing organelle found in phototrophic eukaryotic microorganisms.

chlorosis loss of photosynthetic capability due to bleaching of chlorophyll; symptom of some plant diseases caused by microorganisms.

chlorosomes cylindrical structures containing chlorophyll pigments that are located underneath and attached to the cytoplasmic membrane of some phototrophic bacteria.

cholera disease, characterized by profuse, watery diarrhea that is caused by the ingestion of *Vibrio cholerae*.

C horizon region of the soil beneath the B horizon and above the regolith that contains accumulation of calcium and magnesium carbonates, and is not greatly affected by biological activity.

chromatic aberration defect in a microscopic image due to the refraction of light of varying wavelengths passing through a convex-convex lens; corrected by the use of an achromatic lens; in contrast to spherical aberration.

Chromatium bacterial genus; a purple bacterium that oxidizes reduced sulfur compounds, generally sulfide and thiosulfate, under anaerobic conditions resulting in the formation of sulfate that is deposited with the bacterial cell; intracellular elemental sulfur is used as an electron donor for phototrophic growth when sulfide is absent.

chromatography chemical analysis using a porous solid in which substances are separated by adsorption, fractional extraction, or ion exchange using a variety of techniques such as gas chromatography, high performance liquid chromatography, thin-layer chromatography.

Chromobacterium bacterial genus; gram-negative, facultatively anaerobic, chemoorganotrophic bacilli that are motile by a single polar flagellum and 1-4 subpolar or lateral flagella; produces a bright violet colony when cultured on solid media and a violet ring at the liquid/glass wall of a test tube when grown in nutrient broth; isolated from soil and water; may be pathogenic to humans.

chromogen/chromogenic pigmented; capable of producing color.

chromophore colored portion of a dye.

chromosome a genetic element composed of genes arranged linearly. Most bacteria have one circular chromosome while eukaryotes have multiple linear chromosomes.

chronic continuous, over an extended period of time.

chronic bronchitis disease presenting with productive cough lasting 3 or more months in a year for at least 2 years associated with tobacco smoking and environmental exposure to bioaerosols especially for farmers and workers in agricultural facilities.

chronic health effect long-term adverse reaction.

Chrysonilia sitophila fungal species; produces large numbers of readily aerosolized arthroconidia that may result in contamination of laboratory facilities; colonizes cork stoppers during manufacture and processing, but does not produce 2,4,6-trichloroanisole, guaiacol, or 1-octen-3-ol associated with off flavors of wine stored with corks contaminated with *Penicillium* spp. and other fungi.

C.I. confidence interval.

cidal lethal.

cilia short filaments on eukaryotic cells that beat in sequence resulting in cell motility; same microtubule structure as flagella.

ciliostatic effect a pathological mechanism in the respiratory tract that results in diminished mucociliary clearing and local inflammation in the airway and sinuses; can be elicited by the inhalation route of exposure to mycotoxin.

cilium singular term of cilia.

cinereous ash gray in color.

circinate hooked.

CISH chemiluminescent *in situ* hybridization.

citric acid cycle a cyclical series of chemical reactions within a cell to convert acetate to CO_2 and NADH.

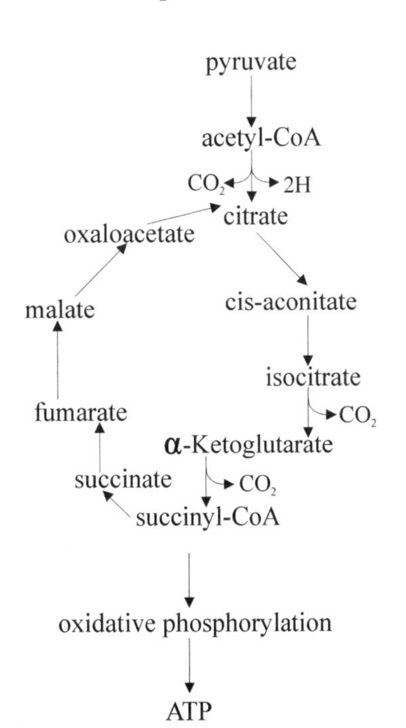

citriform lemon-shaped.

Citrobacter member of the family Enterobacteriaceae; human pathogen with potential environmental exposure via aerosols generated during wastewater treatment practices; member of the total coliform group.

C.L. confidence limit.

Cladosporium fungal genus; slow-growing, olive-brown to blackish brown, velvety colony on malt extract agar that becomes powdery; olive to brown conidiophores and hydrophobic spores that dominate the airborne outdoor population during the day have a wide variation of size and shape with a dark attachment scar and often an olive to brown pigmentation; spores commonly airborne during wet weather and released by water droplets; ubiquitous, currently 40 recognized species; isolated from the soil, plant litter, leaf surfaces, decaying plants, textiles, wood, moist window sills, and refrigerated foods; causative agent of leaf mold of tomato; water activity of 0.85–0.88; potential mycotoxin production of cladosporin.

Cladosporium cladosporioides fungal species; common in soil and plant material and survives drying of soil in summer; conidia of this organism and *Cladosporium herbarium* and among the most common of the airborne fungi.

Cladosporium herbarium fungal species; produces enzymes used in the transformation of steroid intermediates for the industrial production of hormones; formerly termed *Hormodrendron*.

clamp connection a looping bridge connecting the sides of two fungal cells of a hypha that is characteristic of the division Basidiomycota.

clavate club-shaped.

Clavibacter bacterial genus; nonmotile, aerobic gram-positive phytopathogen.

Claviceps purpurea fungal species; parasitic to grains and grasses; produces a variety of mycotoxins including ergotamine, ergotaline, and ergonovine when growing in grain used in preparation of bread.

Clean Water Act a 1977 amendment to the 1972 Federal Water Pollution Control Act which emphasized control of toxic pollutants and established a program to transfer the responsibility of Federal clean water programs to the States; also used as an abbreviated name for the Federal Water Pollution Control Act of 1972.

clearance testing monitoring for airborne and surface-associated microbial contaminants in an indoor environment after completion of remediation activities.

cleistothecia plural of cleistothecium.

cleistothecium a sexual fruiting structure characteristic of the division Ascomycota; type of ascomata that provides for passive spore dispersal as it splits under pressure to release spores in contrast to the active mechanism of a perithecium.

climax community a stable assemblage of microorganisms in a habitat without further succession.

cloned produced as an exact replicate.

clones a group of recombinant DNA molecules, genetically identical cells, or genetically identical organisms that are derived from a single ancestral molecule, cell, or organism.

cloning vector a genetic element that facilitates the recombination and replication of gene sequences; a DNA molecule, such as a plasmid, into which another DNA fragment has been inserted (often from a different source), but which still maintains the capacity for self-replication; used to introduce foreign DNA into a host cell for the purpose of reproducing the DNA in relatively large quantities.

Clostridium bacterial genus; anaerobic, gram-positive endospore-forming bacilli, some of which are aerotolerant.

Clostridium botulinum bacterial species; obligate anaerobic, endospore-forming pathogenic bacterium that is ubiquitous in the environment; pathogenicity associated with the production of neurotoxins A–G with toxins A, B, E, and F associated with food-borne botulism in humans.

Clostridium thermocellum bacterial species; active in the thermophilic biodegradation of cellulose in the soil.

CLPP community level physiological profile.

CLSM confocal laser scanning microscopy.

CM cytoplasmic membrane.

CNS central nervous system.

coagulation the formation of a semisolid mass from a liquid suspension; in drinking water treatment, the joining of dispersed particles and dissolved solids, usually by the addition of a coagulating substance such as alum, resulting in the formation of particles sufficiently large to settle out by gravity.

CO_2 fixation the reduction of carbon dioxide to an organic compound; requires energy and NADH or NADPH.

cocci plural of coccus.

Coccidioides fungal genus; a dimorphic fungus with a mycelial phase and a yeast-like phase; the mycelial phase is characterized by a fast growing, flat smooth or cottony colony having aerial hyphae and alternating arthroconidia that are 2–4 μm × 3–6 μm barrel-shaped cells; spherules are formed in the parasitic, yeast phase.

Coccidioides immitis fungal species; causative agent of coccidiomycosis; endemic to the soil of semiarid regions, notably the southwest desert areas in the United States; infective in the vegetative mycelial phase.

coccidiomycosis disease caused by the inhalation of arthrospores of *Coccidioides* *immitis*; commonly referred to as valley fever; generally reported as a respiratory infection with flu-like symptoms with malaise, fever, pneumonia, and a rash, may progress to severe disease.

coccobacilli/coccobacillus a bacterium whose width is only slightly less than its length.

coccoid sphere-shaped.

coccus a sphere-shaped bacterium.

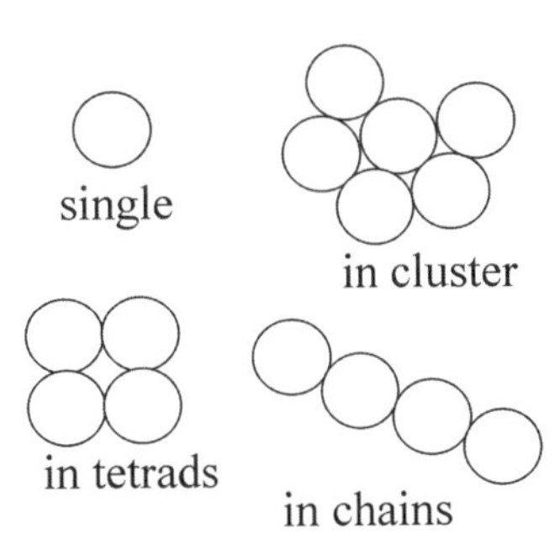

COD chemical oxygen demand.

codon a three base sequence in messenger RNA that encodes for a specific amino acid.

coefficient of determination (R^2) the proportion of variance that is explained by regression; the proportion of variance in the dependent variable that is explained by the independent variable.

coenocytic multinucleated cell.

coenzyme a low molecular weight molecule that accepts or donates electrons during enzymatic reactions.

Cohn, Ferdinand Julius (1828–1893) professor of Botany who devised sterilization by intermittent heating in 1876, one year before Tyndall; described *Bacillus subtilis* endospores and their heat-resistant properties; scientist credited with the first attempt to establish taxonomy of bacteria based on their constant and characteristic morphology.

cohort a subset of a defined population being studied.

cohort study prospective study.

coincidence count mathematical manipulation of data to avoid errors due to the recording of multiple objects as a single entity.

CO_2 incubator an enclosed, temperature-regulated chamber supplied with CO_2 at a specified concentration for the growth of bacteria in culture; also used for the growth of cell culture.

colicin bacteriocin produced by strains of *Escherichia coli*.

coliform/coliform bacteria informal term for total coliform bacteria; gram-negative, aerobic and facultative anaerobic, bacilli that ferment lactose with the formation of gas within 48 hours at 35°C and do not form endospores; includes *Citrobacter* spp., *Klebsiella* spp., *Enterobacter* spp., and *Escherichia coli*.

Colilert a rapid commercially available diagnostic test to detect the presence of total coliform bacteria and *Escherichia coli*.

coliphage a virus that infects coliform bacteria.

collarette a remnant of cell wall that is attached around the tip of a phialide.

collocated sample one of two or more samples collected at the same location and time.

colloid a particle, approximately 1 to 1000 nm in size, that is dispersed within a liquid medium in such a manner as to prevent rapid settling; viruses and clay particles behave as colloids.

colonization the establishment of microbial growth on a surface.

colony the macroscopically visible growth of a microorganism on a culture medium consisting of many cells.

colony-forming unit (CFU) a single cell or spore or cluster of cells or spores that produces a colony on solid medium.

colorimetric using color to measure a specific reaction.

columella the inflated apex of a sporangiophore.

Comamondaceae bacterial family; members have demonstrated the ability to degrade anilines and chloroanilines that are present in the environment as a result of transformation of herbicides.

comb wide-toothed strip used in gel electrophoresis to form wells in the agarose medium.

combustible capable of burning or catching fire.

cometabolism the transformation of a compound by the action of two or more microorganisms, but the organisms are unable to utilize the compound for energy, carbon, or other nutrient.

Commensal/commensalism a unidirectional relationship of microbial populations in which one population benefits and the other population is unaffected; occurs when the metabolic processes of one population alters the environment making it more favorable for the other population, when one population produces and excretes growth factors that are used by the other population, or one population transforms insoluble compounds or converts organic molecules that can then be used by the other population.

commercial off the shelf (COTS) commercially available reagents, test kits, or other laboratory supplies.

commodity chemical inexpensive chemical sold in bulk.

common source a single location.

common vehicle of exposure a nonliving object, such as a food product or water source, that serves to transmit an infectious agent to a group of individuals; a fomite if it is an inanimate object.

community an assemblage of microorganisms that are living together and interacting within a given habitat.

community level physiological profile (CLPP) a rapid, relatively inexpensive method used to gather information about entire communities of microorganisms to allow differentiation among different communities; method consists of inoculating samples into wells containing different substrates and evaluating the pattern of substrate utilization.

community water system a drinking water supply that serves at least 25 year-round residents or has at least 15 service connections used by year-round residents; compare to noncommunity water system.

compatible solutes compounds that are accumulated inside cells to counteract the dehydration of cells in environments with high osmotic concentrations.

compensation depth/compensation zone the lowest level in surface water where light can penetrate and photosynthetic activity balances respiratory activity.

competence property that permits a cell to undergo transformation.

competition a negative relationship between microbial populations in which the survival and growth of both populations are adversely affected by the actions of the other.

competitive exclusion agent (CE agent) bacteria or mixtures of bacterial species that can suppress infection in the host by other microorganisms; in contrast to probiotic.

complementary base pairs pairing of adenine and thymine (A-T) (adenine and uracil in the case of RNA) or guanine and cytosine (G-C) to each other via hydrogen bonds from opposite strands of a double-stranded nucleic acid (DNA or RNA), thereby holding the double-stranded nucleic acid together.

complementary DNA (cDNA) a molecule of DNA that was synthesized from a messenger RNA template.

completed water quality testing the evaluation of samples being tested for the presence of coliform bacteria that were positive during confirmatory water quality testing by Gram reaction and the inoculation into lactose broth-supplied test tubes with a Durham tube and incubation at 35°C; the observation of gas production viewed as trapped gas bubbles in the Durham tube is recorded as a positive result.

complex media/complex medium mixture used for culture of microorganisms in which the chemical composition is unknown; in contrast to defined medium.

compost, composting microbial process that converts organic waste into stable, humus-like material that can be used for soil improvement; decomposition is initiated by mesophilic heterotrophs followed by thermophilic bacteria and fungi as the temperature increases due to biological activity; optimal water content of the material is 50–60% with carbon:nitrogen ratios of 40:1; inhalation risk exposure of workers in composting facilities due to high levels of pathogenic microorganisms.

concatamer long, linear DNA segment comprised of two or more separate molecules that are linked end to end.

concave shaped with an inner curve or depression; in contrast to convex.

concentrate to increase the concentration of a substance by decreasing the unit volume or area.

concentration the amount of a substance in a defined volume, weight, or area.

concentration factor a numerical designation for the increased amount of material retained following a treatment whereby a 10-fold increase is a concentration factor of 10.

concolorous uniform in color.

concrescent growing together.

concurrent cohort study longitudinal study.

concurrent prospective study longitudinal study.

conductivity of water measurement of the concentration of ions present and the temperature at which the measurement is made but does not indicate the nature of the substances in solution; expressed as µmho/cm.

confidence interval (C.I.) a statistical term describing the area above and below the mean that encompasses values that are acceptable as the mean; a 95% confidence interval indicates that values within this area have a 95% likelihood of being a correct and a 5% likelihood of being an incorrect representation of the true value.

confidence limit (C.L.) the upper and lower values of the confidence interval.

confirmatory water quality testing further evaluation of water samples being tested for the presence of coliform bacteria that were positive during presumptive water quality testing by inoculation onto m-Endo agar.

conflict of interest situation where an individual has a vested interest in the outcome of an event.

confluent culture/confluent growth describes the presence of microorganisms on the entire surface area.

confocal scanning laser microscopy (CLSM) technique that uses a light microscope with a laser light source, photomultiplier detectors, and a computerized digital imaging providing optical sectioning of the sample; when combined with fluorescent staining this technique permits the determination of live organisms in a liquid suspension with increased sensitivity and decreased out-of-focus blur.

confounding factor a condition that influences an outcome, but is not the object of the investigation.

conic/conical cone-shaped.

conidiogenous cell the cell that produces a conidium.

conidia plural of conidium.

conidial head referring to the apex of the conidiophore.

conidiophore a stalk-like structure that bears a conidiogenous cell.

conidium asexual spore of genera in the Deuteromycotina; a specialized, non-motile asexual spore; in contrast to a sporangiospore and an ascospore.

conjugation the transfer of genetic elements from one prokaryotic organism to another involving cell-to-cell contact; a sexual process that occurs in bacteria, ciliate protozoa, and certain fungi in which genetic material is exchanged during temporary fusion of two cells.

conjunctiva membrane lining the eyelids and covering the exposed surface of the eyeballs.

conjunctivitis an inflammation of the membrane that covers the outer surface of the eye and the inner surface of the eyelid.

consensus sequence the nucleotide sequence that occurs most frequently at a defined location in a related set of nucleic acids.

consortia plural of consortium.

consortium an association of two or more organisms in which each organism performs a function for the benefit of the other(s).

constant depth film fermenter (CDFF) a device used to study biofilms in which scraper blades pass over pans of various depths resulting in a constant thickness of the biofilm.

contaminant candidate list (CCL) a listing of chemical and biological contaminants under consideration for regulation; established by the United States Environmental Protection Agency.

continuous referring to a fungal structure without septa.

continuous cell line a cell line that will grow essentially indefinitely *in vitro*, given the necessary nutrients and environmental conditions, in contrast to a primary cell culture.

continuous emission monitoring systems (CEMS) instrumentation designed for continuous monitoring of airborne pollutants release from a source.

continuous exposure guidance levels (CEGLs) established for the Department of Defense by the National Research Council Committee on Toxicology as maximum allowable exposure concentrations that are not to be exceeded to avoid immediate or delayed adverse health effects with exposure periods up to 90 days.

continuous feed reactor composting facility that moves decomposing material through a series of stages to produce a more uniform product in a shorter period of time than the aerated pile composting or static pile composting methods.

continuous-flow slide culture (CFSC) a continuous culture device used to study biofilms and other microbial communities; a glass flow cell mounted on a microscope stage and irrigated with a liquid sample provides a means to observe the development of microbial populations over time.

continuous variables data that are not whole numbers but may be expressed within a defined range; in contrast to discrete variables.

control a process designed to determine the validity of a test.

convex shaped with an outer curve or bulge; in contrast to concave.

Coomassie blue a dye used in electrophoresis techniques to stain proteins.

cooperation positive interactions of organisms; demonstrated in the growth curve of a fastidious bacterial culture with a very small inoculum as a long lag period because there are not enough organisms to overcome leakage of low molecular weight metabolic intermediates needed for synthesis and growth; demonstrated in the populations that utilize insoluble substrates by the combined production of extracellular enzymes.

copiotrophic high nutrient concentration; in contrast to oligotrophic.

coprozoite parasite that lives in feces.

copy number the number of molecules of a particular type, such as a plasmid or gene, in a cell.

coremium a bundle of parallel conidiophores.

core polysaccharide one of the three regions of the gram-negative bacterial cell membrane's lipopolysaccharide; less variability in this region compared to that observed for the O-specific polysaccharide.

core sample material collected using a cylindrically shaped device inserted into soil or sediment.

coriaceous leathery in appearance.

corn meal agar (CMA) a growth medium consisting of freshly ground cornmeal, agar and water used for the isolation of fungi.

Cornwall pipettor syringe-shaped device designed to deliver a prescribed amount of liquid.

coronavirus a member of the Coronaviridae family of viruses; large (80–160 nm in diameter), spherical, enveloped, single-stranded RNA viruses with large spikes that project from the envelope; cause respiratory and enteric infections in humans.

correlation coefficient Pearson product moment correlation coefficient.

cortex the inner layer of the endospore spore coat that surrounds the core.

Corynebacterium bacterial genus; characterized as gram-positive, nonmotile, curved bacilli that do not form endospores, but have a swelling that results in a club-shaped appearance; some species are phytopathogens.

Corynebacterium facians bacterial species; phytopathogen, causative agent of leafy gall of ornamentals.

Corynebacterium insidiosum bacterial species; phytopathogen, causative agent of wilt of alfalfa.

Corynebacterium michiganese bacterial species; phytopathogen, causative agent of wilt of tomato/tomato canker.

Corynebacterium sepedonicum bacterial species; phytopathogen, causative agent of ring rot of potatoes.

coryneform club-shaped.

COTS commercial off the shelf.

Coulter counter/Coulter multisizer instrumentation designed for electronic particle counting of microorganisms in solution without regard to viability; generally used to enumerate bacterial cells and fungal spore suspensions for comparison to culture and polymerase chain reaction amplification assay.

coumarins pleasant smelling compounds released by wilting plants that have anticoagulant activity; some mycotoxins belong to this group of compounds.

counterstain dye generally applied following decolorization to contrast in color with a primary stain.

cov covariance.

covariance (cov) the product of deviation between two values from their respective means.

covariate a variable measured that may have an effect on the primary variable of interest.

coverslip thin sheet of glass with multiple uses in the microbiology laboratory such as a cover over a wet mount specimen for viewing with light microscopy;

also used as a collection surface for some impactor samplers (e.g., Air-O-Cell), and as a support matrix for slide culture techniques.

coxsackievirus a member of the enterovirus genus in the Picornaviridae family; small (22–30 nm), icosahedral, nonenveloped, single-stranded RNA viruses; transmitted by the fecal-oral or inhalation route of exposure; have been isolated from sewage, drinking water, and ground water; cause a wide range of illnesses, including aseptic meningitis, upper respiratory illness, acute hemorrhagic conjunctivitis, myocarditis, and infantile diarrhea.

CPE cytopathic effect.

Crapper, Thomas London plumber credited with the invention of the flush toilet in the late 19th century by installing a U-bend siphoning system to clean the collection pan.

Crenarchaeota a kingdom of Archaea representing the hyperthermophiles.

crenulate scalloped.

Crick, Francis presented the model of the structure of DNA with Watson in 1953.

crista the inner membrane of a mitochondrion.

criteria air pollutants group of unwanted airborne chemicals or particulates regulated by the United States Environmental Protection Agency.

critical orifice opening of a defined size.

crossover trial two study populations that differ in their treatment are observed for a defined period and then the treatments are switched and the populations observed for the same period of time.

cross react false positive results obtained in an analysis due to lack of specificity of a reagent.

cross-sectional study an investigation that determines prevalence in a population by measuring the exposure and the outcome at the same time.

cross-tabulation method to establish the cause of illness in a group of individuals by reviewing attack rates when multiple possible sources of the suspected contaminant and other factors are present.

cruciform cross-shaped.

cryofixation fast cooling low temperature preparation of samples for scanning electron microscopy used to prevent the formation of ice crystals in the sample.

cryophile psychrophile.

cryopreservation use of low temperatures for storage of samples.

cryovial small capsule used for the storage of reagents or samples in an ultrafreezer.

Cry proteins/Cry toxins crystalline proteins produced by *Bacillus thuringiensis* that are toxic to insects.

cryptic growth increase in concentration of organisms utilizing nutrients supplied from dead cells of other organisms.

cryptococcosis disease caused by the yeast *Cryptococcus neoformans*.

Cryptococcus fungal genus; an asexual basidiomycete occasionally isolated in indoor environments; previously designated as *Torulopsis*.

Cryptococcus neoformans fungal species; causative agent of cryptococcosis, a meningitis that may also involve the kidneys, liver, prostate, and the lung resulting from the inhalation route of exposure.

cryptosporidiosis diarrheal disease caused by the coccidian protozoan parasite *Cryptosporidium parvum*.

Cryptosporidium parvum protozoan species; chlorine-resistant oocysts are passed in feces from infected animals and humans resulting in the contamination of water sources; numerous outbreaks in drinking and recreational waters have been documented; infection is especially severe in immunocompromised individuals, while immunocompetent individuals generally experience self-limiting diarrhea.

crystal violet stain used in the Gram reaction to impart a non-decolorizing, dark blue to purple coloration to gram-positive bacteria.

CSLM confocal scanning laser microscopy.

CTC a fluorescent compound used to estimate the percentage of active metabolizing cells in a sample.

cubic feet per minute (CFM, ft^3/min) rate measurement of the volume of air or liquid per unit time; 1 CFM = 28 liters/min = 0.028 m^3/min).

culturable the ability of a microorganism to grow under laboratory conditions.

cultural eutrophication process of becoming rich in nutrients as a result of human activity.

culture the growth of a microorganism under laboratory conditions.

culture-based assay the cultivation of a microorganism from a sample.

culture collection archived system of cataloged biological materials.

cumulative frequency polygon variation of the frequency polygon in which the cumulative count of each interval is used instead of the mid-point of the interval.

cumulative percent transforming data to a percentage of the total.

cuneiform wedge-shaped.

Curvularia fungal genus; erect, pigmented, geniculate conidiophores with single conidia that are characterized as having three or more transverse septa with an overall curved shape due to an enlarged central cell; commonly found in dry outdoor air; 35 recognized species common to subtropical and tropical regions where they are phytopathogens resulting in leaf spot diseases.

cutaneous pertaining to the skin.

CVS cardiovascular system.

CWA Clean Water Act.

cyanobacteria diverse and widely distributed group of phototrophic bacteria with >1000 recognized species that use CO_2 as their carbon source and H_2O as their electron donor for oxygenic photosynthesis; under certain conditions these organisms can conduct anoxygenic photosynthesis using H_2S as their electron source; termed blue-green algae or blue-green bacteria.

cycloheximide fungicidal and algacidal agent often used as an amendment to bacteriological culture media to minimize the growth of fungi; trade name Actidione.

cyst resistant, resting stage structure of some bacteria and protozoa.

cytopathic effect (CPE) the deterioration of tissue culture or cell culture cells caused by viral infection; visualization of the deterioration is considered evidence of virus presence in a sample.

Cytophaga bacterial genus; chemolithotrophic members of the Cytophagales that have deep yellow-orange or red pigments and hydrolyze agar, cellulose, and chitin, and degrade organic matter in the environment.

Cytophagales group of gliding bacteria; some genera are chemolithotrophic (e.g., Beggiatoa and Cytophaga).

Cytophaga psychrophila bacterial species; former designation for *Flavobacterium psychrophilum*.

cytoplasm the aqueous material of a cell that is internal to the cytoplasmic membrane; it contains the cellular organelles.

cytoplasmic membrane the selectively permeable layer of the cell that separates the cytoplasm from the environment.

cytosine a pyrimidine base that is one of the four nucleotides comprising DNA and RNA molecules; cytosine forms a bond with guanine on the opposite strand of the DNA molecule.

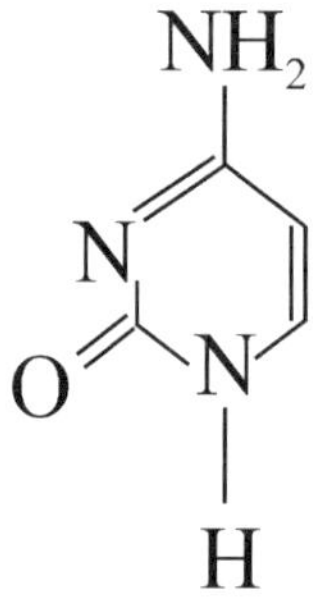

cytotoxic lethal to cells.

Cyt proteins cytolytic proteins produced by *Bacillus thuringiensis* that are toxic to insects.

CZ agar Czapek agar.

Czapek (CZ) agar a fungal culture medium used for the isolation of fungi and bacteria that can utilize inorganic nitrogen.

Czapek Dox agar a variation of Czapek agar fungal culture medium with an increase of sodium nitrate that is used as a reference agar for identification of *Aspergillus* species.

D

d symbol for species richness.

DAPI 4,6-diamidino-2-phenylindole; a DNA-binding fluorochrome used in direct counting with epifluorescent microscopy to determine viability of cells.

dark-field microscopy technique in which light is directed to an unstained specimen only from the sides resulting in a scattering of the light and the visualization of a light object against a dark background; often used to detect motility of microorganisms in wet-mount preparations; in contrast to phase-contrast microscopy.

dark repair mechanism of DNA repair following exposure to ultraviolet light that occurs in the absence of light; also termed excision repair; in contrast to photo repair.

data plural term for the collective numerical or descriptive events recorded as results of an experiment or an observation.

data quality objectives (DQOs) statements that describe the level of uncertainty in the results that is acceptable for regulatory decisions.

data sheet permanent record of results obtained in the laboratory (laboratory data sheet) or activities conducted during sample collection (field data sheet).

dauxie the utilization of one substance at a given rate before the utilization of another substance at a different rate.

DBP disinfection by-product.

DBT dibenzothiophene.

DDBJ DNA Data Bank of Japan.

DDT 1,1,1-trichlorobis(*p*-chlorophenyl) ethane, a biodegradation-resistant pesticide due to the presence of *p*-chloro substitutions; in contrast to methoxychlor.

death phase final period of a growth curve of a microbial culture in which the number of organisms decreases over time.

de Bary, Anton German botanist who proved experimentally in 1861 that *Phytophthora infestans* was the causative agent of potato blight giving rise to the science of plant pathology.

decant to pour off a liquid.

decimal reduction time the time required to inactivate 90% of the microorganisms in a sample at a specified temperature; estimated resistance of a microbial cell or endospore to a particular process at a specified temperature; commonly termed D value.

decolorization quenching or destaining of a dye.

deconvolution the sharpening or deblurring of an image obtained with an optical microscope by removal of the out-of-focus image.

decussate arranged in pairs at right angles to the subsequent pair.

deep agar test tube with medium solidified without a slant that is generally inoculated with a needle in a single stab.

defined media/defined medium mixture used for the culture of microorganisms in which the chemical composition of all of the ingredients are known; in contrast to complex media.

definitive host the organism in which the sexual cycle of a parasite occurs.

degeneracy a condition in which multiple codons encode for the same amino acid.

dehiscence separating, splitting, or tearing away with a circular action; used to describe the liberation of conidia.

deionized water (DI water) water that has been treated with an ion exchanger to remove ions.

delayed-incubation fecal coliform procedure a modification of the fecal coliform membrane filter technique to be used when a field incubator is not available; samples are filtered onto M-ST holding medium to keep fecal coliform bacteria viable, but prevent them from growing during transport.

delayed-incubation total coliform procedure a modification of the standard total coliform membrane filter method to permit the transport of samples after filtration; samples are filtered onto M-ST holding medium to keep total coliform bacteria viable, but prevent them from growing during transport.

delimiting establishing the boundaries between two substances or areas.

deliquescing liquefying or dissolving away at maturity.

de Man-Rogosa-Sharpe (MRS) medium culture medium for growth of lactobacilli often used in the study of natural fermentation processes.

dematiaceous descriptive of the dark pigmentation of some fungal hyphae and conidia.

dematiaceous fungi group of deeply pigmented fungi that are found in soil or on decaying organic materials.

denaturation the separation of double-stranded DNA into two single strands by manipulation of the ionic conditions of the solution; in contrast to melting for separation; in contrast to hybridization for the construction; also refers to the breaking of hydrogen bonds to alter tertiary structure of proteins.

denaturing gradient gel electrophoresis (DGGE) a gel electrophoresis method used to separate DNA fragments of the same length, but containing different base-pair sequences; used to determine presence and abundance of different microbial species in a mixed population.

dendroid tree-like in appearance.

denitrification the conversion of nitrate to nitrogen gas under anoxic conditions; a dissimilatory nitrate reduction reaction.

density ratio of mass to volume, often expressed in grams/cubic centimeter.

density gradient technique that uses a centrifuge to separate particles based on density.

denticle small, narrow tooth-like outgrowth that supports conidia.

denticulate finely-toothed.

deoxynivalenol mycotoxin; also called vomitoxin; may cause laboratory changes in immunoglobulins with exposure via handling of contaminated foodstuffs.

deoxyribonucleic acid (DNA) a double-stranded molecule that encodes genetic information in cells; the molecule is composed of sequences of four nucleotide bases: adenine (A), guanine (G), thymine (T), and cytosine (C); the two strands of the molecule are held together by weak bonds between A—T and G—C base pairs; the sugar is deoxyribose making it stable to basic solutions; in contrast to ribonucleic acid.

depauperate appearing starved or underdeveloped.

dependent variable the outcome of interest that changes in response to an intervention.

depth of field the distance between the closest and farthest objects in focus within a field of view of a lens.

dermal pertaining to the skin.

dermal route of exposure a situation in which foreign material is absorbed or penetrates through the skin.

dermatitis inflammation of the connective tissue underlying the epidermis of the skin; possible reaction to exposure to some mycotoxins.

Dermatophilaceae bacterial family; taxonomically in the order Actinomycetales; characterized as saprophtyes or facultative parasites that produce spores that are not borne in sporangia and have mycelia that divide transversely to form motile cocci; representative genera include *Dermatiophilus* and *Geodermatophilus*.

descriptive statistics used to describe data; the presentation, organization, and summarization of data such as geometric mean, median, and standard deviation.

DeSaussure scientist who demonstrated in 1839 that soil microorganisms were involved with the oxidation of hydrogen gas in the soil, an activity that was eliminated when the soil was heated, or treated with a salt solution or an acid.

desiccant a material that absorbs moisture and is used to maintain reagents in a dry state.

desiccate/desiccation drying.

desiccator jar a storage vessel containing a desiccant.

Desulfacinum bacterial genus; group II acetate-oxidizing dissimilatory sulfate reducing bacteria that are nutritionally diverse, thermophilic, gram-negative cocci to oval-shaped cells that utilize C1-C18 fatty acids and are also capable of autotrophic growth.

Desulfitobacterium dehalogenans bacterial species; gram-positive halorespiring bacterium that links the oxidation of hydrogen, formate, lactate, and pyruvate to the reduction of organic and inorganic acceptors including ortho-chlorinated phenols.

Desulfoarculus bacterial genus; group II acetate-oxidizing dissimilatory sulfate reducing bacteria that are motile, gram-negative vibrios that utilize C1–C18 fatty acids as electron donors.

Desulfobacter bacterial genus; group II dissimilatory sulfate reducing bacteria that are gram-negative bacilli; 4 currently recognized species that utilize only acetate as the electron donor to oxidize CO_2 via the citric acid cycle.

Desulfobacterium bacterial genus; group II acetate-oxidizing dissimilatory sulfate reducing bacteria; 3 species of marine organisms that are capable of autotrophic growth via the acetyl-CoA pathway.

Desulfobacula bacterial genus; group I non-acetate-oxidizing dissimilatory sulfate reducing bacteria; one marine species of oval to coccoid cells that can oxidize various aromatic compounds such as toluene to CO_2.

Desulfobotulus bacterial genus; gram-negative, group I non-acetate-oxidizing dissimilatory sulfate reducing bacteria; 1 currently recognized species.

Desulfobulbus bacterial genus; group I non-acetate-oxidizing dissimilatory sulfate reducing bacteria that are gram-negative with ovoid-shaped cells; 3 currently recognized species.

Desulfococcus bacterial genus; group II acetate-oxidizing dissimilatory sulfate reducing bacteria that are gram-negative, nonmotile cocci that utilize C1–C14 fatty acids as electron donors with oxidization to CO_2 and are capable of autotrophic growth via the acetyl-CoA pathway.

Desulfomicrobium bacterial genus; group I non-acetate-oxidizing dissimilatory sulfate reducing bacteria that are motile, gram-negative bacilli; 2 currently recognized species.

Desulfomonile bacterial genus; group I non-acetate-oxidizing dissimilatory sulfate reducing bacteria that are capable of reductive dechlorination of 3-chlorobenzoate to benzoate.

Desulfonema bacterial genus; group II acetate-oxidizing dissimilatory sulfate reducing bacteria that are large, gram-positive, non-spore-forming organisms that utilize C2–C12 fatty acids as electron donors with complete oxidation to CO_2 and are capable of autotrophic growth via the acetyl-CoA pathway with H_2 as the electron donor.

Desulforhabdus bacterial genus; group II acetate-oxidizing dissimilatory sulfate reducing bacteria that are nonmotile, non-spore-forming, gram-negative bacilli that utilize fatty acids with complete oxidation to CO_2.

Desulfosarcina bacterial genus; group II acetate-oxidizing dissimilary sulfate reducing bacteria that are gram-negative, non-spore-forming cells arranged in packets that utilize C2-C12 fatty acids as electron donors with complete oxidation to CO_2 and are also capable of autotrophic growth via the acetyl-CoA pathway with H_2 as the electron donor.

Desulfotomaculum bacterial genus; group I non-acetate-oxidizing dissimilatory sulfate reducing bacteria that are motile, gram-negative bacilli; 4 currently recognized species, one of which is thermophilic and one that is capable of utilizing acetate as the energy source.

Desulfovibrio bacterial genus; group I non-acetate-oxidizing dissimilatory sulfate reducing bacteria that are gram-negative, motile, curved bacilli with 12 recognized species, one of which is thermophilic.

Desulfuromonas bacterial genus; obligate anaerobic gram-negative bacilli that are dissimilatory sulfur reducing bacteria that can utilize acetate, succinate, ethanol, and propanol as the electron donor; these organisms couple the oxidation of acetate and ethanol to the reduction of elemental sulfur to H_2S but they are not sulfite reducers; 4 species currently recognized.

determinate a conidiogenous cell that ceases to grow before or after the formation of the first conidium; in contrast to indeterminate.

deterministic model term used to describe a mathematical model in which the input variables, and consequently the output information, are considered to have single values; in contrast to probabilistic model.

Deuteromycetes Deuteromycotina.

Deuteromycotina a subdivision of the Amastigomycota; grouping of saprophytic, symbiotic, parasitic, or predatory fungi that are characterized as unicellular or more typically with septate mycelium, usually producing conidia from various types of conidiogenous cells; no sexual stage has been recognized; more than 15000 species of fungi in this group also known as Fungi Imperfecti.

Deutsche Sammlung von Mikroorganismen und Zellkulturen GmbH (DSM) culture collection entity located in Braunschweig, Germany that catalogs and sells standard strains of microorganisms.

DFA direct fluorescent antibody staining method.

DG18 dichloran-18% glycerol-agar.

DGGE denaturing gradient gel electrophoresis.

dH_2O distilled water.

d'Herelle, F. described bacterial viruses as bacteriophages in 1917 independently from Twort.

DI water deionized water.

dialysis technique in which molecules are separated by size exclusion through a semiporous membrane.

diarrheal food poisoning an illness that is characterized by abdominal pain with diarrhea 8–16 hours after ingestion of contaminated food; associated with ingestion of a variety of pathogens such as *Bacillus cereus*-contaminated foods including meats and vegetable dishes, pasta, desserts, cakes, sauces, milk, and water.

diatom an algal cell that has a cell wall composed of hydrated silica embedded in an organic matrix.

diatomaceous earth soil composed of deposits of diatoms.

diauxic growth/diauxie the growth of a microorganism in culture with two nutrient compounds, one compound is utilized first and then the second compound is utilized, resulting in a biphasic growth curve.

diazotroph an organism that is capable of nitrogen fixation.

dibenzothiophene (DBT) used as a model compound to demonstrate microbial desulfurization.

DIC differential interference contrast microscopy.

dichloran-18% glycerol-agar (DG18) a growth medium used for the isolation of moderately xerophilic fungi.

dichotomous branching into two segments; a property or characteristic having only two possible outcomes—it is either observed or it is not observed.

dichotomous key sequential arrangement of properties used for taxonomic classification of microorganisms proceeding from the general to the specific.

difference imagery a technique that utilizes the selective differences between two digital images.

differential culture medium a reagent-amended preparation that provides a means to visually differentiate between microbial populations.

differential interference contrast (DIC) microscopy light microscopic technique in which subtle differences in cell structure are observed; polarized light passes through a prism generating two separate beams that pass through the specimen prior to entering the objective lens where they are combined into one; the image is the result of the differences in the refractive index of the material that the two beams pass through; also termed Nomarsky.

differentiation the distinguishing of one feature from another; modification of cellular structure or function.

diffluent readily dissolving.

diffraction the change in the direction of a wave caused by its encountering an object.

diffusion movement from a source to an area of lower concentration, no energy is required for this process.

DIG digoxigenin.

digoxigenin (DIG) a nonfluorescent compound used as a marker for *in situ* hybridization experiments.

diluent liquid that is used to decrease the concentration of a substance in solution.

dimethyl sulfoxide (DMSO) produced by photooxidation of dimethyl sulfide and during the degradation of phytoplankton in marine environments; used as a terminal electron acceptor by some bacteria and fungi in freshwater habitats; used as a solvent to transport substances that are not readily water soluble across cell membranes.

dimorphic having two forms; diphasic.

diphasic descriptive of fungi that have two distinct morphologic forms (filamentous or yeast-like) depending on the temperature during incubation.

diphasic fungi characterization of some fungal genera in which there is a filamentous phase and a yeast phase.

diphtheroid shaped like a diphtheria bacillus.

direct causation a result is linked to one factor without an intermediate step or an additional factor; in contrast to indirect causation.

direct count enumeration of microorganisms using methods that are not reliant on viability but measure the total number present.

direct fluorescent antibody staining technique in which the antibody is fluorescent; in contrast to indirect fluorescent antibody staining.

direct transmission transfer of a pathogen from an infected host without an intermediate; in contrast to indirect transmission.

disarticulation breaking apart.

disc gel a gel in which the chemical composition (e.g., pH, buffer concentration) is not present as a continuous gradient; often run in a cylindrical tube so that the bands appear disk-like.

discoid disk-shaped.

discrete variables data that have only a limited set of values; data that are whole numbers; in contrast to continuous variables.

disinfect to treat a surface or a liquid with a substance that will kill microorganisms.

disinfectant a chemical used on surfaces or in water to kill microorganisms; may cause harm to host tissue.

disinfection by-product (DBP) any one of a number of chemical compounds that are formed by the reaction of organic compounds in water or wastewater with the disinfectant (e.g., chlorine), the resultant chemicals may be carcinogenic; examples of DBP include the trihalomethanes, haloacetic acids, chloral hydrate, haloketones, haloacetonitriles.

disjunctor a connective cell or cell wall material found between the conidia of some fungal species.

dispersal transmission of microorganisms from a source, generally via air movement, water splash, or mechanical disruption.

dispersion the spread of a substance from the source; in statistics it is the measurement of how closely the data cluster around a typical value or central tendency.

disphotic zone area aphotic zone.

dissimilative/dissimilatory sulfur/sulfate reduction the conversion of SO_4^{-2} to H_2S, but the resulting sulfur is not utilized for biosynthesis; in contrast to assimilative sulfur reduction.

dissimilatory nitrate reduction the reduction of nitrate (NO^{3-}) to nitrite (NO^{2-}), which is then further reduced to either nitrogen gas (N_2) or nitrous oxide (N_2O); nitrate is used by the microorganisms as a terminal electron acceptor; the reduction of nitrate to nitrogen gas under anoxic conditions is termed denitrification.

dissimilatory sulfate reducing bacteria currently 18 recognized genera of obligate anaerobic bacteria that are categorized into 2 groups; group I (e.g., *Desulfobulbus*, *Desulfomicrobium*, *Desulfotomaculum*, and *Desulfovibrio*) is comprised of non-acetate oxidizing organisms that utilize lactate, pyruvate, ethanol, or some fatty acids as carbon and energy sources to reduce sulfate to H_2S; group II (e.g., *Desulfobacter*, *Desulfococcus*, *Desulfonema*, *Desulfosarcina*) is comprised of acetate oxidizing organisms that reduce sulfate to sulfide.

dissimilatory sulfur reducing bacteria genera (e.g., *Desulfuromonas*) that are capable of reducing elemental sulfur to sulfide but are not able to reduce sulfate to sulfide; in contrast to chemoorganotrophic faculative aerobic bacteria that reduce elemental sulfur and thiosulfate, sulfite, and DMSO (e.g.,

Proteus, *Pseudomonas*, *Campylobacter*, and *Salmonella*).

distal away from the point of origin.

distilled water (dH_2O) water that has been purified by boiling and recondensing of the steam in a clean vessel, thereby removing nonvolatile constituents.

distoseptate to be separated from; used to describe multicellular conidia in which the individual cells have a thick secondary wall that is separate from the outer wall of the conidium.

diurnal repeated daily or occurring during the day.

divaricate highly divergent.

division taxonomic level in the classification of bacteria below domain, includes several orders.

DMSO dimethyl sulfoxide.

DNA deoxyribonucleic acid.

DNA chip DNA microarray.

DNA Data Bank of Japan (DDBJ) the sole DNA data bank in Japan officially certified to collect DNA sequences from researchers and to issue the internationally recognized accession numbers; produced collaboratively with GenBank and the EMBL Nucleotide Sequence Database.

DNA fingerprinting molecular technique to determine the relationships among organisms.

DNA microarray molecular method for detection of specific microorganisms in which hundreds to thousands of oligonucleotide probes are fixed onto a support matrix, labeled sample DNA is added, hybridization occurs, and the identity of the complimentary sequences is determined.

DNA polymerase an enzyme that directs the synthesis of DNA by the addition of nucleotides in the order prescribed by a template.

DOC dissolved organic carbon.

doliform jar-shaped.

DOM dissolved organic matter.

domain highest taxonomic level in the classification scheme, includes several kingdoms.

dormant in a resting stage, not metabolically active.

dorsiventral having a front and a back.

dose the amount of a substance, generally that ingested or applied at one time.

dose rate the amount of a substance ingested or applied per unit time.

dose response assessment the area of risk assessment that encompasses the determination of the relationship between the magnitude of exposure and the probability of an adverse effect.

dot blot a method used to detect and/or quantify a specific nucleic acid sequence; the solution being analyzed is spotted onto a matrix (such as a membrane filter), the probe is applied and allowed to hybridize to the sequence on the matrix; an estimate of the quantity present is possible if various dilutions of the solution are analyzed and the results compared to spots of known concentrations.

dot plot presentation of data that is similar to a bar chart, except dots are used in place of the bar; also termed point graph.

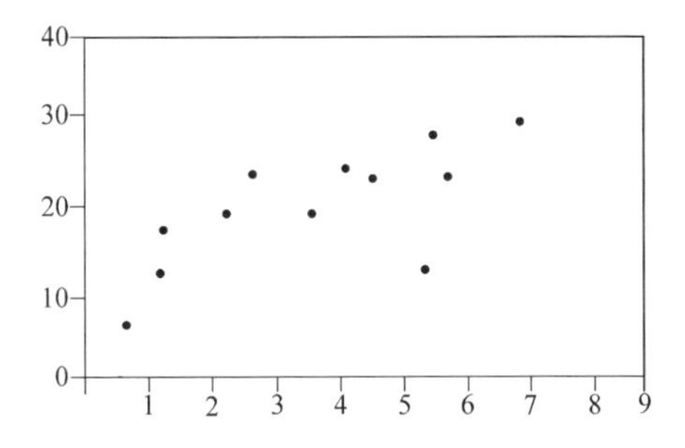

double-agar overlay a standard method used for the detection and enumeration of bacteriophage in which the sample and bacterial host are added to molten soft agar that is then poured over hardened agar; the plate containing the double agar is then inverted and incubated to allow for the development of visible plaques.

double blind investigation in which neither the investigators nor the subjects know which individuals belong to the test group or the control group.

doubling time generation time.

downstream primer the primer that binds to the 3′ end of the target sequence.

downy mildew disease of plants caused by fungi that are characterized by a light colored fluffy growth on leaves and stems.

DQOs data quality objectives.

Dreshlera fungal genus; velvety to wooly textured colony on malt extract agar that is white in color becoming olive brown on the surface and reverse; brown simple or branched conidiophores are geniculate and poroconidia are fusoid, distoseptate and do not have a protuberant hilum which distinguish this genus from *Exserohilum*; germ tubes are produced from any cell which distinguish this genus from *Bipolaris*; isolated from soil and plants, and many species are phytopathogens; colonies readily become sterile in laboratory culture.

drinking water standard municipal drinking water standard.

dry air spora airborne fungal spores that are passively distributed from the site of colonization due to warm dry weather; often the most abundant airborne fungi in the environment; generally the concentration of these spores is maximal during the afternoon hours when humidity is low and wind speeds are increased; airborne concentrations are also increased during initial rainfall events as spores are released from leaf surfaces by raindrops.

dry rot a condition of powdery, dry decomposition of timber in which the composition of the wood is changed by fungal activity, but the wood has not been subjected to water damage or moisture accumulation; generally associated with the activity of basidiomycetes in structural timber when the water necessary for fungal metabolism is not present at the site of the damage, but is transported from the soil through microstructures or is condensed onto fine surface fibrils of the fungus.

dsDNA double-stranded deoxyribonucleic acid.

dsRNA double-stranded ribonucleic acid.

Durham tube a small glass tube inserted upside down into a large test tube that contains lactose growth medium for presumptive water quality testing, complete water quality testing, and the Eijkman test; the small inverted tube is designed to trap gas bubbles as an indicator of gas production generated as a result of fermentation of the lactose by the fecal coliform bacteria present in water samples.

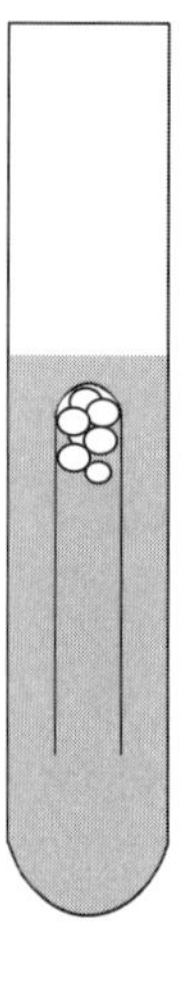

dust fine, solid particles.

dust sample/dust sampling settled dust sample/settled dust sampling.

D value decimal reduction time.

dysgonic poorly growing.

dyspnea labored or difficult breathing.

E

E_h redox potential.

Eastern equine encephalitis (EEE) virus a member of the Alphavirus genus (also referred to as group A arboviruses) of the Togaviridae family; 60–70 nm diameter, icosahedral, single-stranded RNA viruses; transmitted by mosquitoes; this is the most severe of the group A arboviruses, with mortality in 50–75% of infected individuals; see also Western equine encephalitis virus, Venezuelan equine encephalitis virus.

eccentric asymmetrical in growth.

echinate/echinulate spiny.

echovirus a member of the enterovirus genus in the Picornaviridae family; small (22–30 nm), icosahedral, nonenveloped, single-stranded RNA viruses; transmitted by the fecal-oral or inhalation route of exposure; have been isolated from sewage, drinking water, and ground water; cause a wide range of illnesses, including aseptic meningitis, respiratory illness, encephalitis, and diarrhea.

E. coli *Escherichia coli*.

E. coli C a strain of *Escherichia coli* that is used as a host for somatic coliphage.

E. coli C3000 a strain of *Escherichia coli* that is used as a host for somatic coliphage.

E. coli CN-13 a nalidixic acid mutant of *E. coli* C that is used as a host for somatic coliphage.

***E. coli* Famp** a strain of *Escherichia coli* used as a host for male-specific (F + RNA) coliphages; contains the Famp plasmid, and is resistant to ampicillin and streptomycin.

***E. coli* K-12** early *Escherichia coli* isolate used for genetic manipulations.

***E. coli* O157:H7** bacterial strain that causes hemorrhagic colitis, hemolytic uremic syndrome, and thrombocytopenia resulting from an ingestion route of exposure to contaminated food, such as undercooked beef and raw milk, or to fecal-contaminated drinking or recreational water.

ecology the study of the relationships between organisms and the environment; derived from the Greek *oikos* (dwelling) and *logos* (law).

ecosystem a community of organisms in their natural environment.

ectendomycorrhizae combination of ectomycorrhizae and endomycorrhizae.

ectomycorrhizae association in the mycorrhizae in which the fungus forms an external sheath and the fungal hyphae penetrate the intercellular spaces of the root but do not invade living cells of the plant; in contrast to endomycorrhizae.

ectoparasite organism that lives on or within the skin.

ectopic misplaced, occurring in an abnormal place.

edema accumulation of fluid in the tissue.

EDS energy dispersive X-ray spectroscopy.

EDTA ethylenediaminetetraacetic acid.

EDX energy dispersive X-ray analysis.

EEE Eastern equine encephalitis virus.

EEGLs emergency exposure guideline levels.

EELs emergency exposure limits.

efferent leaning away from; in contrast to afferent.

EFM epifluorescent microscopy.

EHEC enterohemorrhagic *Escherichia coli*.

EHS extremely hazardous substance.

Eijkman test simplified protocol to eliminate the three-step presumptive water quality testing, confirmatory water quality testing, and completed water quality testing by inoculation of dilution of water samples into lactose broth and incubation at 44.5°C resulting in the growth of fecal coliform bacteria as indicated by the generation of gas bubbles trapped in a Durham tube and the inhibition of nonfecal coliforms due to the elevated temperature.

Eimeria protozoan species; intracellular occcidian of vertebrates and invertebrates.

electron acceptor the compound that is reduced (i.e., gains electrons) during a chemical or biochemical reaction.

electron donor the compound that is oxidized (i.e., loses electrons) during a chemical or biochemical reaction.

electronic particle counting a method whereby an instrument is used to enumerate the particles of various sizes in a liquid.

electronegative having a negative electrical charge; of or pertaining to a chemical element that becomes an anion by taking on electrons.

electron microprobe and wavelength-dispersive spectroscopy (EM/WDS) an analytical spectroscopy technique that offers better resolution than energy dispersive X-ray spectroscopy and provides quantitative assessment of elemental composition of the sample.

electron microscopy the use of electrons and electromagnets in a vacuum instead of light beams and lenses to study detailed structure of microorganisms.

electrophoresis technique of separation of ionic molecules according to their size or their charge based on their migration through a gel to which an electric field has been applied; smaller molecules with a more negative charge will move more rapidly and further toward the anode; bands of molecules may be visualized by staining, such as with ethidium bromide or Coomassie blue.

electroporation method for introduction of plasmids into a host cell in which an electrical pulse is used to make the cell membrane permeable.

electropositive having a positive electrical charge; of or pertaining to a chemical element that becomes a cation by losing electrons.

electrorotation diagnostic technique in which microorganisms are subjected to an uniform rotating electric field; the organisms rotate in a characteristic manner depending on the conductivity and permittivity of the cell wall, plasma membrane, and cytoplasm, and the suspending material.

electrospray ionization mass spectrometry (ES-MS) analytical chemistry detection method for the determination of biological molecules following separation by high performance liquid chromatography.

ELISA enzyme-linked immunosorbent assay.

ELSD evaporative light scattering detection.

eluate the liquid suspension resulting from elution.

elute/elution the process of dislodging of material retained on a filter matrix using a small volume of liquid; the separation of one material from another using a liquid in which one of the materials is soluble but the other is not.

EMB Eosin methylene blue agar.

Embden-Meyerhof-Parnas pathway a series of biochemical reactions in which glucose is ultimately converted to pyruvate with the production of ATP.

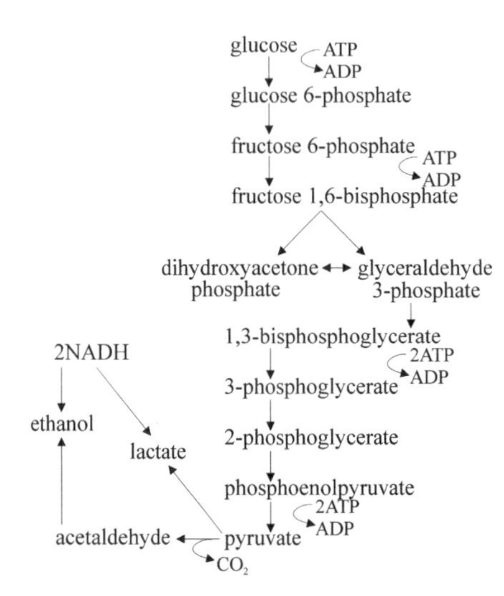

EMBL Nucleotide Sequence Database a European database that contains DNA and RNA sequences; produced collaboratively with GenBank and the DNA Data Bank of Japan.

emergency exposure guideline levels (EEGLs) developed by the National Research Council for use by the US Department of Defense to provide guidance for the exposure of military personnel to hazardous chemicals while operating under emergency conditions for which other agencies have not set relevant standards.

emergency exposure limits (EELs) concentrations of atmospheric contaminants that can be inhaled for single periods over a defined period of time without causing irreversible toxicity but not without causing some irritation or systemic intoxication.

emergency response planning guidelines (ERPG) values in a three tiered rating system published by the American Industrial Hygiene Association to assist environmental and health and safety professionals develop emergency response strategies for the protection of workers and the general public exposed to chemicals and other substances.

emergency shower safety device in the laboratory that provides a large volume of water in a cascade to rapidly rinse a chemical contaminant from an exposed individual.

emerging pathogen newly recognized or documented disease-causing microorganism.

Emericella fungal genera; teleomorph of some *Aspergillus* species.

emetic food poisoning an illness characterized by nausea and vomiting 1–5 hours after ingestion of contaminated food; associated with a variety of *Bacillus cereus*-contaminated foods predominantly oriental rice dishes.

emigration the movement of a population out of an environment; in contrast to immigration.

EMS ethyl methane sulfonate.

EM/WDS electron microprobe and wavelength-dispersive spectroscopy.

encephalitis an inflammation of the brain.

electropositive having a positive electric charge.

encrusted covered with a layer of mineral.

encyst to form a cyst.

endemic present in a given population or geographic area at all times.

endemic pathogen disease-causing microorganism that is present in a given population or geographic area at all times.

endergonic requires the input of energy; in contrast to exergonic.

endobiotic within a living body; in contrast to epibiotic.

endocytosis the envelopment of a particle within a cell.

endogenous developed or living within an organism.

endolithic microorganisms that live within the rock matrix; in contrast to lithobiotic.

endomisity the natural population of microorganisms in a specified area.

endomism the biogeography of microorganisms.

endomycorrhizae association in the mycorrhizae in which the fungus invades living cells of the plant root; in contrast to ectomycorrhizae.

endoparasite parasite that lives within the body of another organism.

endophyte an organism that lives within a plant; generally used to describe endomycorrhizae.

endospore heat resistant structure formed by some bacterial genera that confers resistance to environmental stresses of dryness, heat, and reduction of nutrients needed for growth; produced as the protoplasm of the cell is reduced to a minimum volume as the result of accumulation of Ca^{2+} and the synthesis of dipicolinic acid forming a gel-like substance followed by the formation of a thick cortex around the protoplast core; spores of *Coccidioides immitis* that are formed within a spherule.

endosymbiosis/endosymbionts micro-organisms that live within the cell of another organism in a mutualistic relationship.

endothermic a reaction that requires energy; in contrast to exothermic.

endotoxin heat-stable lipopolysaccharide lipid A complex of the outer membrane of gram-negative bacilli; released into the environment during cell lysis or active growth; increased toxicity observed when intact cells are phagocytized by macrophages; present in dusts as microvesicles with concentrations remaining stable for long periods of time; inhalation route of exposure related to agriculture activities, dusty environments, and humidified air handling systems; affect the humoral and cellular host mediation systems; may result in structural and functional changes in lung tissue following respiratory exposure with reports of chest tightness, cough, shortness of breath, fever and wheezing in occupational settings and has been associated with atopic and non-atopic asthma in humans; associated with organic toxic dust syndrome; analyzed using *Limulus* amebocyte lysate assay.

end-point analysis a procedure in which results are obtained only at the completion of an assay without intermediate data points or measurements recorded.

energy dispersive X-ray spectroscopy (EDS)/energy dispersive X-ray analysis (EDX) an analytical scanning electron microscopic technique in which the wavelength of X-rays produced by electron-specimen interaction are used to characterize the chemical composition of a sample.

enhanced surface water treatment rule (ESWTR) a rule proposed by the United States Environmental Protection Agency to increase protection of finished drinking water supplies from contamination by *Cryptosporidium* and other pathogens; it will apply to public water systems using surface water or ground water under the direct influence of surface water.

enrichment culture the use of selective conditions to favor the growth of a particular microorganism by permitting the organism to outgrow other organisms, inhibiting the growth of other organisms, or distinguishing the organism from other organisms in the sample.

enteric intestinal.

enteric cytopathic human orphan virus echovirus.

enteric virus any one of many viruses that infect the mammalian gastrointestinal tract; generally excreted in fecal material and therefore have the potential for waterborne transmission.

Enterobacter bacterial genus, member of the family Enterobacteriaceae; facultatively anaerobic, straight, gram-negative bacilli that are motile by peritrichous flagella; environmental isolates grow better at 20–30°C while clinical isolates grow better at 37°C; widely distributed in nature.

Enterobacter agglomerans bacterial species; human pathogen with potential environmental exposure via aerosols generated during wastewater treatment practices; inhabits the gut of termites where it can fix nitrogen; found in soil; phytopathogen that causes dry necroses, galls, wilts, and soft roots.

Enterobacter cloacea bacterial species; most frequently isolated *Enterobacter* species from humans and animals; human opportunistic pathogen isolated from respiratory tract and wounds, but not the feces; potential environmental exposure via aerosols generated during wastewater treatment practices, but not an enteric pathogen.

Enterobacteriaceae bacterial family; characterized as gram-negative bacilli that are either non-motile or motile by means of peritrichous flagella.

Enterococcus bacterial genus; gram-positive homofermentative cocci in chains that are primarily associated with the intestine and feces of mammals; members of a subgroup of the fecal streptococci; generally considered to be better indicators of fecal pollution than fecal streptococci; formerly classified as *Streptococcus*.

Enterococcus faecalis bacterial species; saprophytic commensal that inhabits the intestine and oral cavity of humans and animals; used as an indicator of fecal pollution in water; can serve as an opportunistic pathogen.

Enterococcus faecium bacterial species; shown to be resistant to newly designed antibiotics due to exposure to closely related antibiotics present in animal feed.

Enterolert commercially available diagnostic test for the detection of enterococci.

enterotoxin a toxin that affects the cells of the intestinal tract.

enterovirus viral genus; of the family *Picornaviridae* consisting of more than 100 mammalian and insect viruses; normal habitat is in the intestine of mammals; in mammalian systems, primary site of infection is the gastrointestinal tract; infection may spread to other sites including the nervous system; poliovirus is the prototype enterovirus, other enteroviruses of interest in mammalian systems include the coxsackieviruses and the echoviruses.

entomopathogenic capable of causing disease in insects.

envelope surrounding material; a membranous structure composed of lipids, proteins, and carbohydrates that surrounds some types of viruses; may have surface projections, spikes, composed of glycoproteins.

enveloped virus a virus with an envelope surrounding the nucleocapsid.

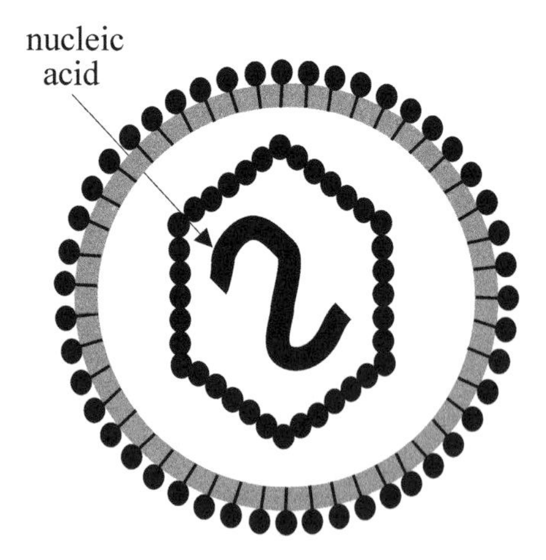

Envirochek caspule filter a 1 μm nominal pore size pleated polyether sulfone filter in a polycarbonate housing used in the US Environmental Protection Agency method 1622 for the concentration of *Cryptosporidium* oocysts from water samples.

environmental chamber self-contained unit used for incubation of specimens or for the conduct of experiments under controlled conditions.

environmental microbiology the study of microorganisms that grow in or contaminate the environment.

Environmental Protection Agency (EPA) generally used designation for the United States Environmental Protection Agency, but also used to denote a local municipality or state agency with activities varying by legislative mandate.

enzootic a disease present in an animal population at all times.

enzyme-linked immunosorbent assay (ELISA) immunoassay method that uses enzymes to increase the sensitivity of antibody detection.

Eosin Methylene Blue (EMB) agar a selective culture medium for the isolation of gram-negative bacteria that minimizes the growth of gram-positive bacteria and also serves as a differential culture medium by incorporating eosin and methylene blue to distinguish between lactose and non-lactose fermenting organisms; medium specified in the membrane filtration method for analysis of drinking water for coliform bacteria; first developed by Holt-Harris and Tague.

EPA Environmental Protection Agency; United States Environmental Protection Agency.

EPCRA emergency planning and community right to know.

epibiotic on a living body; in contrast to endobiotic.

Epicoccum fungal genus; ubiquitous, two currently recognized species; colony on malt extract agar is felt-like to wooly in texture and yellow, orange, red, or brown in color with a deep brown reverse and a orange or brown pigment diffusing into the medium; distinctive attached spores released by an active spore discharge associated with hygroscopic movement; isolated from plant debris paper, textiles, and insects; water activity of 0.86–0.90; associated with Type I allergies; produces antibiotic substances.

epidemic the occurrence of an unusually high number of individuals in a localized area infected with a specific pathogen, with the number in excess of that normally observed in a localized region.

epidemiologist a scientist that studies the occurrence, distribution, and control of infectious diseases in populations.

epidemiology the study of the cause, occurrence, distribution, and control of infectious diseases and other health-related effects or events in specified populations.

epifluorescent microscopy fluorescent microscopy in which the light is transmitted through the objective onto the specimen rather than directed through the specimen.

epigenesis theory that an organism develops by gradual increase in complexity.

epilimnion upper thermal zone of water that is typically warm and oxygen-rich during the summer but the photosynthetic activity depletes mineral nutrients, located above the thermocline.

epipelagic zone designation for the vertical region in the marine ecosphere that is generally euphotic and warm and ranges from 0–200 meters in depth.

epiphytic descriptive of microorganisms that inhabit plant stems, leaves, and fruit.

epispore the outer coat of a fungal spore.

epithecium the layer of tissue surrounding a fungal spore sac.

epizootic the occurrence of an unusually high number of animals in a localized region infected by a particular microorganism; an epidemic among animals.

EPS extracellular polysaccharides, exopolysaccharides.

equilibrium condition in which the concentration is not changing.

equitability (J) the component of species diversity that measures the proportion of individuals among the species present, thereby indicating if there are dominant populations; this component is independent of sample size and can be calculated from the Shannon-Weaver Index of Diversity; in contrast to species richness.

$$J = \frac{\text{mean Shannon-Weaver diversity index}}{\text{theoretical maximal Shannon-Weaver diversity index}}$$

ergosterol a sterol present in the fungal cell wall; a component that can be used to estimate the amount of fungal biomass present in a sample.

ergot purplish black mycotoxin-containing sclerotia formed by *Claviceps purpurea* when growing on grains and grasses.

ergotism food poisoning caused by ingestion of mycotoxin-contaminated grain which resulted in symptoms of bizarre behavior; prevalent during the 14th to 18th centuries.

Erlenmeyer flask a laboratory flask that is shaped with a broad base and straight sloping sides.

ERPG emergency response planning guidelines.

ERPG-1 the first tier in the emergency response planning guidelines that is the maximum airborne concentration below which it is believed that nearly all individuals could be exposed for up to 1 hour without experiencing more than mild transient adverse health effects or without perceiving a clearly defined objectionable odor; in contrast to ERPG-2 and ERPG-3.

ERPG-2 the second tier in the emergency response planning guidelines that is the maximum airborne concentration below which it is believed that nearly all individuals could be exposed for up to 1 hour without experiencing or developing irreversible or other serious health effects or symptoms that could impair an individual's ability to take protective action; in contrast to ERPG-1 and ERPG-3.

ERPG-3 the third tier in the emergency response planning guidelines that is the maximum airborne concentration below which it is believed that nearly all individuals could be exposed for up to 1 hour without experiencing or developing life threatening health effects; in contrast to ERPG-1 and ERPG-2.

erumpent breaking out.

Erwinia bacterial genus; member of the Enterobacteriaceae; facultatively anaerobic, epiphytic, straight bacilli, all but one species are motile by peritrichous flagella; species are phytopathogens others are saprophytic.

Erwinia amylovora bacterial species; phytopathogen, causative agent of fire blight of fruit trees, especially pear and apple.

Erwinia cartovora bacterial species; functions as cloud condensation nuclei when airborne; phytopathogen, causative agent of soft rot of fruit, black leg of potato, and blight on the chrysanthemum.

Erwinia herbicola bacterial species; spoilage organism of fresh fruit.

Erwinia stewartii bacterial species; phytopathogen, causative agent of wilt of corn.

erythema redness of the skin.

Escherichia bacterial genus; member of the family Enterobacteriaceae; gram-negative, facultatively anaerobic, straight, bacilli that are motile by peritrichous flagella or nonmotile; isolated from the lower intestinal tract of warm-blooded animals.

Escherichia coli bacterial species; common inhabitant of the lower bowel of humans and animals; used as an indicator organism in assays of water and food.

ES-MS electrospray ionization mass spectrometry.

Esp enterococcal surface protein.

ESWTR enhanced surface water treatment rule.

estuarine of, or pertaining to, an estuary.

estuary a coastal area where mixing of fresh water and marine waters occurs, has high productivity.

ethanol (EtOH) an organic chemical compound frequently used as a surface disinfectant.

ethanologenic produces ethanol as a by-product of metabolism.

ethidium bromide a fluorescent dye used to label double-stranded DNA and viewed with an ultraviolet light.

ethyl alcohol ethanol.

ethylenediaminetetraacetic acid (EDTA) a chelating agent added to samples to reduce metal toxicity.

ethylene oxide a chemosterilant gas that acts as an akylating agent to sterilize temperature-sensitive materials such as plastics and medical instruments; rarely used for foods due to residual taste.

etiologic agent a microorganism that causes disease.

etiologic fallacy ascribing characteristics of a population to individual members who do not have those characteristics.

etiology the cause or origin of a disease.

EtOH ethanol.

EU unit of measurement for endotoxin.

eubacteria bacteria.

eugonic luxuriant growth.

Eukarya a phylogenetic classification of organisms containing the fungi, animals, plants, algae, slime molds, entamoebae, ciliates, flagellates, trichomonads, microsporidia, and diplomonads.

eukaryote a microorganism that is usually >2–100 μm in size with linear DNA that is enclosed in a membrane-bound nucleus; in contrast to prokaryote.

eukaryotic flagella a thread-like extension of the ectoplasm that is part of the neuromotor apparatus of some eukaryotic microorganisms; in contrast to the flagellum of prokaryotic microorganisms.

euphotic zone area of effective light penetration to the compensation zone in the marine ecosphere; combination of the littoral and limnetic zones in lake water where photosynthetic activity occurs.

eutrophication process of becoming rich in nutrients; occurs when the balance of Liebig's Law of the Minimum is disturbed by the input of limiting nutrients.

Eurotium fungal genus; teleomorph, cleistothecial stage of *Aspergillus glaucus* group.

Euryarchaeota a kingdom of Archaea representing methanogens, extreme halophiles, and the genus *Thermoplasma*.

eurytopic/eurytolerant an organism with a wide distribution or the ability to tolerate a wide range of environmental conditions.

eutrophic having a high nutrient content.

evanescent disappearing rapidly.

evaporation the conversion of liquid to a vapor.

evolutionary distance the separation between branches on a phylogenetic tree as inversely proportional to relatedness.

excision repair dark repair.

exergonic the release of energy; in contrast to endergonic.

exfoliation peeling or flaking of the skin.

exine outer wall of a fungal spore.

exogenous arising on the exterior or outside.

exon nucleic acid sequence of eukaryotic genes that forms the finished mRNA molecule; in contrast to intron.

Exophiala fungal genus; yeast-like mucoid colony on malt extract agar that becomes velvety to slightly floccose in texture, dark green to dark brownish green, brown or black in color with septate hyphae that are pale brown; conidia are unicellular or bicellular with an oval to cylindrical shape, hyaline or pale brown in color that accumulate at the tip of the annellides and often appear as if sliding down the side; isolated from soil, plants, animals and humid environments; pathogenic isolates grow at 37°C while saprophytes do not.

exopolysaccharides extracellular polysaccharides.

exothermic a reaction that releases energy; in contrast to endothermic.

exotoxin a toxin that is released outside of the cell; in contrast to endotoxin.

experiment a procedure or series of procedures conducted to test a hypothesis.

experimental design details of the procedures and methods used during the conduct of a research project.

exponential growth/exponential phase the period of time in a growth curve of a microbial population in which the number of cells doubles during each unit of time; presented graphically as a constantly increasing slope.

exposure contact.

exposure assessment the area of risk assessment that encompasses the determination of the extent of contact with a defined agent that may result in an adverse effect.

exposure route the manner in which contact is made between an agent and a host.

Exserohilum fungal genus; phytopathogen that is also present in soil; colony on malt extract agar is velvety in texture with a surface and reverse that are dark olive to black in color; the hyphae are pale brown and septate with brown conidiophores that are bent at the apex; conidia have a protuberant hilum and many species exhibit end cells with a dark end septum; in contrast to *Bipolaris* and *Dreshlera*.

extension a term used in molecular biology to refer to the addition of nucleotide bases to the primer during the polymerase chain reaction amplification process.

extracellular polysaccharides (EPS) heat stable, water-soluble, nonbranched glycoproteins.

extrachromosomal outside of a chromosome; may be used to refer to a gene encoded by a plasmid.

extramatrical outside of the substrate; in contrast to intramatrical.

extreme halophile organism that requires >10% NaCl for growth; isolated from solar salt evaporative ponds, natural salt lakes, and highly salted foods.

extremely hazardous substance (EHS) term defined by the United States Environmental Protection Agency in the Superfund Amendment and Reauthorization Act Title III.

extrinsic allergic alveolitis hypersensitivity pneumonitis.

exudate liquid released from within a source.

eyewash station a place in the laboratory where emergency equipment for flushing contaminants from the eyes and face is located.

F

F– term used to describe bacteria that lack the F plasmid; recipient cell for F+ in bacterial conjugation.

F+ term used to describe bacteria that contain the F plasmid, which carries the genetic information for the synthesis of the sex pilus; sex pilus is site of attachment for male-specific bacteriophage; donor cell.

FI an F+RNA phage that belongs to serological group IV of the RNA phages.

f2 an F+RNA phage that belongs to serological group I of the RNA phages.

face velocity the average velocity (e.g., in feet/minute) of air passing through a defined area of a chemical fume hood or biological safety cabinet; used to determine if the safety apparatus is operating effectively.

FACS fluorescent-activated cell sorting.

factorial analysis of variance/factorial ANOVA statistical analysis to compare multiple groups in a single test and consider interactions between the factors; in contrast to a one-way ANOVA.

factorial design the use of a large study population or sample matrix to test two independent treatments.

	treatment b +	treatment b −
treatment a +	both a and b	a only
treatment a −	b only	neither a nor b

facultative capable of growth in the presence or absence of an environmental factor.

facultative anaerobe/facultatively anaerobic/facultative aerobe/facultatively aerobic a microorganism that is capable of growth with or without oxygen.

facultative parasite an organism that can exist either as a parasite or as a saprophyte.

FADC fluorescent antibody direct count.

false negative result that incorrectly identifies the absence of the analyte of interest when it is actually present in the sample; in contrast to a false positive.

false positive result that incorrectly identifies the presence of the analyte of interest when it is absent from the sample; in contrast to a true positive.

FAME analysis fatty acid methyl ester analysis.

family taxonomic classification of bacteria below order, containing several genera; suffix -aceae denotes family level; highest taxonomic level generally used when describing prokaryotes.

fascicle/fasciculate gathered in bundles.

fastidious having very specific nutritional and environmental requirements.

fatty acid methyl ester (FAME) analysis analytical method used to characterize the fatty acids present in the lipids of bacterial membranes; cells are cultured under specific temperature and nutrient conditions, and then the harvested fatty acids are hydrolyzed, extracted, and derivatized to form methyl esters that are analyzed with gas chromatography and the pattern of peaks is compared to a database for identification of the cells.

FCM flow cytometry measurement.

fd a member of the Inoviridae family of viruses; a male-specific single-stranded DNA bacteriophage.

fecal coliform bacteria aerobic and facultative anaerobic, gram-negative, bacilli-shaped, thermotolerant coliform bacteria that do not form endospores and ferment lactose with the production of gas within 48 hours at 44.5°C.

fecal coliform membrane filter technique analytical procedure to detect fecal coliform bacteria using the membrane filter technique; membranes are placed on M-FC culture medium; fecal coliform bacteria produce blue colonies after 24 hours of incubation at 44.5°C.

fecal-oral route of transmission refers to the transmission of enteric organisms via the ingestion of water or food that has been contaminated with fecal material.

fecal streptococci members of several genera of *Enterococcus* (formerly classified as *Streptococcus*), including *Enterococcus faecalis*, *E. faecium*, *E. avium*, *E. bovis*, *E. equinus*, and *E. gallinarum*; inhabitants of the gastrointestinal tract of humans and other warm-blooded animals; used as indicators of fecal contamination.

Federal Water Pollution Control Act a United States of America federal law passed in 1972 that made it illegal to discharge pollutants into waters without a permit, provided federal funding to construct sewage treatment plants, and directed states to set standards for waters other than interstate navigable waters and to establish programs to protect wetlands.

ferment/fermentation/fermentative catabolic ATP-producing reactions in which organic compounds are the primary electron donor and the ultimate electron acceptor.

fermenter a microorganism that uses organic compounds as both primary electron donor and the ultimate electron acceptor.

fermentor a large growth vessel that is used to culture microorganisms, often for industrial or commercial uses.

ferruginous rust-colored.

fetal bovine serum (FBS) serum derived from fetal bovine used as a rich source of nutrients in growth and maintenance media for animal cells growing in culture.

fetal calf serum (FCS) identical to fetal bovine serum.

Fibrobacter bacterial genus; many species are present in the complex of microorganisms isolated from the rumen.

filamentous thread-like.

filamentous phage a bacterial virus in the Inovirus genus.

filiform thread-like.

film mulching use of plastic sheeting on agricultural fields to minimize growth of weeds and conserve soil moisture.

filter concentration the use of a membrane to accumulate material from a relatively large volume of liquid or air that is then washed off with a smaller volume of liquid.

filter feeding a means of ingestion of suspended planktonic microorganisms by sessile benthic and planktonic invertebrates created by maintaining a flow of water with cilia or modified appendages through specialized gills, tentacles or mucous nets; in contrast to grazing.

filter sterilization the use of a filter to remove microorganisms from a liquid, generally conducted with a vacuum pump to draw the liquid through the filter.

filter tower cylindrical device used to hold the filter and the liquid sample.

filtration sampling collection of particles from the air or a liquid by passage through a porous medium, usually a membrane filter; efficiency of collection is dependent on the physical properties of the particles (e.g., size, shape, density), the filter pore size, and the flow rate of the sample through the filter

fimbriae straight filamentous structures of protein that project from the surface of bacteria that are involved in adherence to surfaces, but unlike flagella are not involved in motility.

fingerprinting generic term used to describe identification strategies that catagorize organisms into groups based on similar and distinct characteristics.

fire blanket a safety wrap used in the laboratory to extinguish flame, generally used to extinguish fire on an individual.

first-order epidemic mesoscale spread of a plant pathogen that occurs in a single growing season in a single field or extends to segments of an entire continent.

fixed factor statistical term to denote that a condition either exists or it does not; in contrast to a random factor.

flaccid limber, soft.

flagella plural of flagellum.

flagellum/prokaryotic flagellum a single coiled tube of protein that when rotating will propel a bacterial cell through a liquid matrix; location and number of flagella on a bacterial cell are used in identification.

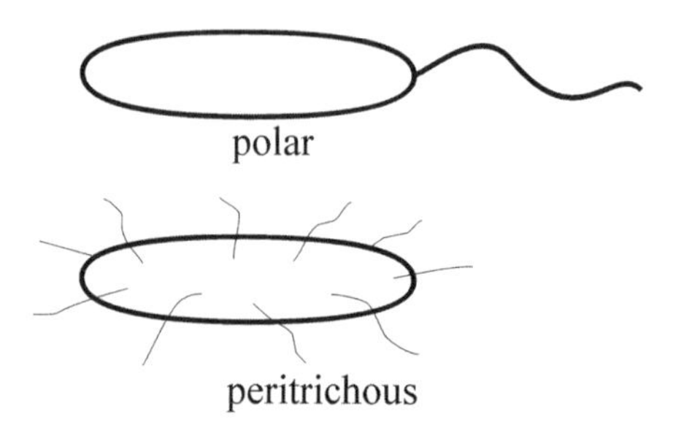

flame sterilization rapid method using an open flame to sterilize metal or glass tools; generally the object to be sterilized is first dipped into ethanol and then passed through the flame although inoculating loops/inoculating needles are generally placed directly into the flame without ethanol.

Flavobacterium bacterial genus, nonmotile, aerobic, gram-negative bacilli with rounded ends; frequently isolated from terrestrial and aquatic environments; active in the carbon cycle due to their ability to degrade a variety of polysaccharides including cellulose, agar, starch, pectin and chitin; several species may be opportunistic pathogens with potential environmental exposure via aerosols generated during wastewater treatment practices; some species are proteolytic and able to degrade casein and gelatin.

Flavobacterium psychrophilum bacterial species; yellow-pigmented, psychrophilic organism that grows at 4–20°C; causative agent of cold water disease in juvenile salmonid, commonly occurring in hatcheries in the spring and winter; formerly termed *Flexibacter psychropilus* or *Cytophaga psychrophila*.

Fleming, Alexander (1881–1955) Scottish bacteriologist and discoverer of antibiotics in 1929 due to the recognition of a zone of inhibition caused by *Penicillium notatum* growing as a contaminant on a culture of *Staphylococcus aureus*.

Flexibacter psychrophila bacterial species; former designation for *Flavobacterium psychrophilum*.

flexuous alternately bent in opposite directions.

FLIPR fluorescence imaging plate reader.

floc a mass formed in a fluid as a result of aggregation of suspended particles;

formed during water treatment after the addition of a coagulating substance such as alum, which causes colloidal materials to aggregate upon contacting one another, allowing them to be removed by gravity settling.

floccose having tufts that are cottony or wooly in appearance.

flocculation a process that causes flocs to form.

flow cytometry measurement (FCM) detection technology that uses simultaneous measurement of light scatter to determine cell size and structure; may incorporate fluorescence to increase capabilities to include quantitation of antigen, and cellular components.

flow-through system chemostat.

fluorescein isothiocyanate (FITC) an organic compound that fluoresces yellow-green; used for the direct counting of microbial cells.

fluorescence exclusion negative staining.

fluorescence imaging plate reader (FLIPR) high-throughput method for analysis of 96-well microtiter plates.

fluorescent having the ability to emit light of a certain wavelength when activated by another wavelength of light.

fluorescent-activated cell sorting (FACS) a flow cytometer technique that utilizes a laser beam to activate fluorescent antibody-labeled cells resulting in a charge to the cells that are then deflected through an electric field and counted.

fluorescent antibody an immunoglobulin coupled with a fluorescent molecule.

fluorescent antibody direct count (FADC) use of an antibody linked to a fluorescent dye for direct count enumeration and identification of microorganisms that offers high degree of specificity provided by the antibody but may also encounter autofluorescence problems due to background material in the sample.

fluorescent *in situ* hybridization (FISH) a microscopic method that can be used for cells in culture or environmental samples with a fluorescent probe specific for cellular ribosomes applied to a sample that has been treated to permit the probe to penetrate the cell membrane and bind to the ribosomal RNA.

fluorescent microscopy microscopic technique that uses excitation of light for the examination of stained specimens.

fluorescent pseudomonads fluorescent members of the bacterial genus *Pseudomonas*; colonize in the rhizosphere and participate in suppression of soilborne diseases.

fluorescent staining use of fluorescent dyes to provide a means for detection.

fluorimeter instrument that measures fluorescence.

fluorite lens an objective lens with fluorite incorporated in the glass of the lens that corrects for chromatic aberration for two colors and spherical aberration for two colors; useful for phase microscopy.

fluorogenic 5′-nuclease assay TaqMan PCR assay.

^{19}F nuclear magnetic resonance (^{19}F NMR) a nuclear magnetic resonance technique used to characterize the degradation of fluorine-containing aromatic compounds by microorganisms.

focal spread distribution of a pathogen from the point of initial infection to surrounding areas.

fomite inanimate object that transmits microbial contaminants to a host.

food chain transfer of energy in organic compounds from one organism to another in a series of steps.

food infection the transmission of an infectious organism through the ingestion route of exposure.

food poisoning a disease resulting from the ingestion of toxic chemicals or toxins produced by microorganisms.

Food Safety and Inspection Service (FSIS) the public health agency of the US Department of Agriculture that is responsible for ensuring the safety of the commercial supply of poultry, meat and egg products and that these products are correctly labeled and packaged.

food web interrelationships of steps in food chains.

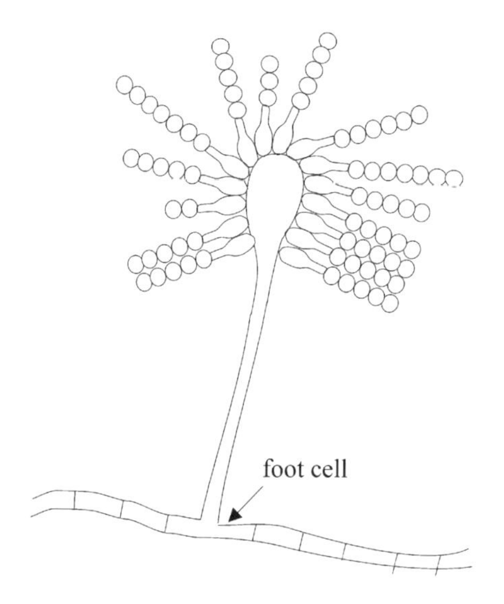

foot cell the base of the conidiophore in *Aspergillus* that is shaped like a foot; characteristic of members of the genus *Aspergillus*.

Foray 48B a *Bacillus thuringiensis* subspecies *kurstaki*-containing biological insecticide.

forced air flow sampler instrumentation for the collection of bioaerosols in which a measured volume of air or a calibrated rate of air flow is sampled; this type of device permits the quantitative measurement of the concentration of airborne microorganisms.

formaldehyde chemical fixative and antibacterial agent.

F-pilus another term for sex pilus; site of attachment of F+ (male-specific) RNA and DNA phage.

F plasmid a plasmid that codes for the production of a sex pilus.

Francisella tularensis bacterial species; gram-negative, obligate aerobic, nonmotile bacillus that requires cysteine for growth; causative agent of tularemia in humans and animals.

Frankia bacterial genus; capable of nitrogen fixation.

Frankiaceae bacterial family; taxonomically in the order Actinomycetales; characterized as free-living in the soil and symbionts in plant nodules; a representative genus is *Frankia*.

free-living does not require a living host, but can live in the environment.

freeze drying lyophilization.

freeze-thaw a method used to cause the release of viruses from infected cells growing in culture; cells are frozen then thawed, usually multiple times to ensure complete release of viruses; also used as an extraction methodology to release nucleic acids from cells for molecular biology techniques (e.g., polymerase chain reaction amplification).

French press the use of high pressure (≥20,000 psi) and extrusion through a narrow orifice to disrupt gram-negative bacterial cells in liquid suspension.

Freon an organic chemical compound used to extract viruses from cell culture lysates.

frequency polygon a diagram of a frequency distribution where the variable is represented by a single dot at the mid point of its value on the x-axis and the frequency is represented as a single dot at the mid point of its value on the y-axis and a polygon is formed by lines.

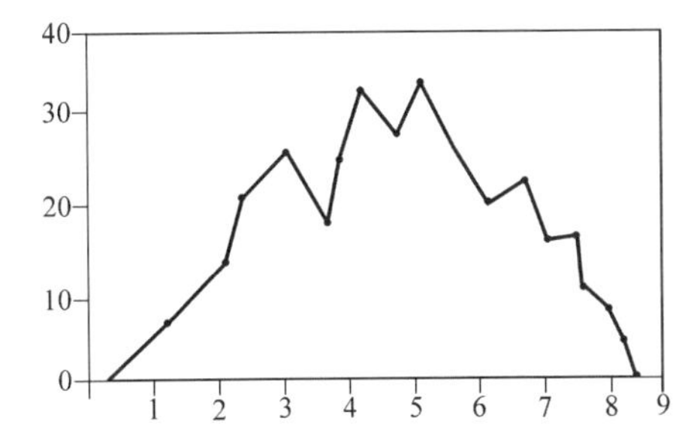

FRhK4 cells cells from the kidneys of fetal rhesus monkeys that have been adapted to grow *in vitro* as a continuous cell line; used for the detection and cultivation of enteric viruses, including hepatitis A virus.

Freund's adjuvant a mixture of killed microorganisms in an oil/water emulsion that induces antibody formation with a greater response than that observed with the organisms alone due to the prolonged exposure that occurs with the slow absorption as a result of the oil.

fructans polysaccharides consisting of fructose units linked by glycosidic bonds; produced in bacteria as part of extracellular polysaccharide where it protects cells from desiccation and aids in adherence to surfaces; present in the cell wall of fungi.

fruiting body a macroscopic reproductive structure with a distinctive shape, size and color produced by some fungi, bacteria, and other organisms.

FSIS Food Safety and Inspection Service.

fugacious disappearing early.

fuliginous smoky or sooty.

fulvous tawny in color.

fungi plural for fungus.

fungicidal an agent capable of killing fungi.

Fungi Imperfecti Deuteromycotina.

fungistatic capable of inhibiting the growth of, but not killing fungi.

fungus eukaryotic, heterotrophic organisms that are either achlorophyllous, saprophytic, or parasitic with filamentous vegetative structures usually surrounded by cell walls composed of chitin or other polysaccharides, propagating with spores and generally exhibiting asexual and sexual reproduction, or unicellular structures; classification is currently primarily based on physiological characteristics.

funiculose having a rope-like or bundled appearance.

furfuraceous having a scruffy appearance; bran-like scales.

Fusarium fungal genus; mycotoxin producer that is ubiquitous; approximately 50–70 species; colony on malt extract agar is pink, orange to purple with soluble pigment on reverse; spores are produced in a slimy mass that is dispersed by insects, water splash and wind; distinctive macroconidia but microconidia may be confused with those of *Acremonium*; isolated from soil, as a saprophyte or pathogen of plants, notably causing root rot, and indoors on very wet surfaces; water activity of 0.86–0.91; associated with Type I allergies, systemic opportunistic infections in disabled people and mycotoxin production; production of zearalenone.

Fusarium culmorum fungal species; mycotoxin producer; aerial mycelium floccose in texture becoming felt-like with age; initially white to yellow or pink becoming orchid to brownish red in color with red, purple, or brown reverse; branched conidiophores with short wide phialides with sickle-shaped macroconidia borne on aerial mycelia; production of trichothecenes and zearalenone.

Fusarium graminearum fungal species; conidia shown to contain T-2 toxin and deoxynivalenol mycotoxins.

Fusarium sporotrichioides fungal species; fast-growing white to yellow, pink, red/purple or brown cottony colony on malt extract agar with branched conidiophores; mycotoxin producer, notably T-2 toxin and deoxynivalenol, with increased production shown in low temperatures and dark conditions.

fusiform spindle-shaped.

fusoid spindle-shaped with two tapered, pointed ends.

G

g gram

GA an F+RNA phage that is the type strain for serological group II of the RNA phages.

Gaeumannomyces graminis* var. *tritici fungal species; causative agent of take-all disease of wheat and barley.

***β*-D-galactosidase** an enzyme produced by coliform bacteria to break down lactose; used in a chromogenic assay because a color change occurs when it hydrolyzes a substrate such as o-nitrophenyl-β-galactopyranoside.

gall a tumorous growth on infected plants; symptom of plant disease caused by *Agrobacterium* spp.

gamma irradiation the use of high energy ionizing radiation to sterilize heat-sensitive materials such as plastics and some foods.

gas chromatography (GC) a separation technique for qualitative and quantitative analysis of organic compounds that can be vaporized before they decompose; sample is introduced into a gas carrier that flows over either a solid (gas-solid chromatography, GSC) or a liquid (gas-liquid chromatography, GLC) stationary phase and is detected by a variety of detectors including flame ionization detector (FID), photoionization (PID), and mass spectrometry (MS).

gastroenteritis an inflammation of the lining of the stomach and intestines resulting in symptoms such as nausea, vomiting, diarrhea, fever; caused by a variety of enteric microorganisms such as Norwalk viruses, *Salmonella* spp., and rotavirus.

gastrointestinal tract mouth, esophagus, stomach, intestines, and related organs.

gastromycetes a grouping of Basidiomycetes that includes puffballs, earth stars, stinkhorns, and bird's nest fungi; spores are not actively discharged.

gas vesicles spindle-shaped, hollow but rigid protein structures present in some aquatic bacterial cells that are used to provide buoyancy and a means of vertical locomotion.

Gaussian distribution presentation of data where the arithmetic mean, median, and mode all have the same value; a data curve that is symmetric around the mean and the tails of the curve get closer to the x-axis the further they are from the mean; also referred to as a normal curve or bell curve; named after the discoverer, Karl Friedrich Gauss.

gauze pad method early method developed for the concentration of viruses from water samples relying on the absorption of the virus to suspended woven material that resulted in a low percent recovery.

gel an inert polymer, generally composed of agarose or polyacrylamide.

gel electrophoresis a method for the separation of nucleic acids and proteins suspended in a gel matrix; the nucleic acids migrate according to their molecular weight and electrical charge for a defined period of time through a gel of an established pore size suspended in a buffer and an electric field.

Gelman filter a specific brand of membrane filters and filter holders used to detect and enumerate bacteria in the membrane filter technique.

gel membrane filtration matrix comprised of gelatinous material that is placed on an agar surface or liquefied during analysis of the retained material.

GEMs genetically-engineered microorganisms.

GenBank the National Institutes of Health (NIH) genetic sequence database, which is an annotated collection of all publicly available DNA sequences; produced collaboratively with the EMBL Nucleotide Sequence Database and the DNA Data Bank of Japan.

gene the basic unit of heredity, consists of one or more segments of DNA required to produce a single polypeptide.

gene library collection of cloned DNA sequences that comprises the genetic information from a single organism.

genera plural of genus.

generally recognized as safe (GRAS) status categorization for microorganisms that are used in the production of food products.

general transduction a process whereby DNA from any portion of the host cell genome becomes part of the DNA of the virus particle, replacing some of the virus genes and rendering the virus defective; the bacterial genes are lost if homologous recombination with the recipient bacterial chromosome fails to occur; in contrast to specialized transduction.

genetically engineered microorganisms (GEMs) microorganisms with *in vitro* altered gene sequences to enhance a particular function.

genetic engineering the *in vitro* manipulation of gene sequences.

genetic recombination recombination.

generation time (t_d) the amount of time required for the number of organisms to double, also termed doubling time; quantified as $t_d = t/n = 1/k$ where t is the time, n is the number of generations, and k is a constant that depends on genetic characteristics of the organism, inoculum size and history, nutrient medium composition, and growth conditions.

geniculate descriptive of a fungal conidiophore that is bent, resembling a knee joint.

genome the complete set of genes in an organism.

geophilic a nonpathogenic fungus that grows in soil rather than on animals or humans.

genotype the genetic composition of an organism; in contrast to phenotype.

genotyping characterizing a microorganism based on the gene sequencing; the use of genetic techniques to identify a genetic defect in an organism.

genus taxonomic classification of microorganisms that contains several species; as a Latin or Greek derivation the term is written in italics or underlined and is generally abbreviated to the first letter followed by a period for subsequent citations within the same document; in bacterial taxonomy it is below the family level.

geometric mean (GM, π) statistical term used to represent the typical value for exponential growth values when the data are not normally distributed but they increase with a sharp rise.

$$GM = \sqrt[n]{\prod_{i=1}^{n} X_i}$$

Geotrichum fungal genus; saphrophyte that is isolated from soil, plants and milk products; may be pathogenic in debilitated individuals; colony on malt extract agar is creamy in texture becoming powdery to waxy that is white in color on the surface and reverse; hyaline hyphae are septate and there are no conidiophores or blastoconidia; rectangular arthroconidia produced with age are released by fission of double walls; occasionally has aerial mycelia resembling *Coccidioides immitis*.

Geotrichum candidum fungal species; colonizes surface-ripened cheeses during the early stages of ripening imparting appearance and aroma characteristics and promoting flavor development.

germ informal term used to describe a microorganism that is capable of causing disease.

germicide/germicidal a substance that inhibits or kills microorganisms.

germination the second stage of germination of endospores following activation that involves the loss of refractility of the spores, increased susceptibility to dye staining, and loss of heat and chemical resistance.

germ slit a narrow, thin-walled groove formed in the wall of a fungal conidium or ascospore where hyphae emerge.

GFP green fluorescent protein.

Giardia lamblia protozoan; thick-walled oval-shaped cyst, approximately 6–8 μm by 10–12 μm, is resistant to environmental conditions and is readily transmitted by the fecal-oral route of transmission from contaminated water, food and fomites; cysts attach to the intestinal wall of mammals and germinate to form vegetative trophozoites, approximately 8–10 μm by 14 μm, that are motile by eukaryotic flagella and block transfer of lipids across the intestinal lining resulting in giardiasis.

giardiasis foul-smelling diarrhea and cramping caused by infection with *Giardia lamblia*.

glabrous smooth.

GLC gas-liquid chromatography; gas chromatography.

Gleophyllum trabeum fungal species; a member of the basidiomycetes; a brown rot fungus.

gliding bacteria grouping of Myxobacterales and Cytophagales based on their motility on solid surfaces.

Gliocladium fungal genus; saprophyte that is isolated from soil and decaying plant materials; colony on malt extract agar is wooly in texture with a white, pink or green color on the surface and a reverse of pale yellow; septate, hyaline hyphae with branched conidiophores and phialides arranged in branch-like clusters with slimy, smooth walled, unicellular conidia arranged in a mass at the apices.

gliotoxin mycotoxin produced by *Gliocladium*.

globose spherical.

glove box air-tight enclosure for the manipulation of samples and cultures under anaerobic conditions.

GLP good laboratory practice.

β-(1→3)-D-glucan/1,3-beta-D-glucan a polyglucose polymer that is a structural component of fungal cell walls; associated with sensory irritation and inflammation-regulating capacity of airway macrophages.

β-glucosidase an enzyme in the cellulase enzyme system that catalyzes the formation of glucose from cellobiose and small oligomers.

gluconeogenesis production of glucose from non-carbohydrates; members of the Archaea accomplish this by reversal of glycolysis using the Embden-Meyerhof-Parnas pathway.

Gluconobacter bacterial genus; acid tolerant, aerobic, gram-negative acetic acid bacilli that ferment ethanol to acetic acid; commonly used in the commercial production of vinegar.

Gluconacetobacter diazotrophicus bacterial species; a strict aerobic, nitrogen-fixing endophyte originally isolated from the root and stems of sugar cane; formerly termed *Acetobacter diazotrophicus*.

glutaraldehyde a chemical fixative used for electron microscopy specimens that interacts with amino groups forming cross-links between proteins.

glycocalyx a polysaccharide-containing material on the external surfaces of a cell that may be arranged as a rigid capsule or an easily deformed slime layer; plays an important role in biofilms, as it aids in trapping nutrients and adherence to the solid matrix.

glycogen a glucose polymer stored in prokaryotic organisms as inclusion bodies.

glycolysis the process of oxidizing glucose to pyruvate.

gm gram.

GM geometric mean.

GMOs genetically-modified organisms; genetically engineered microorganisms

gnotobiotic an environment in which all the biological populations are defined or known; also termed axenic.

good laboratory practice (GLP) a system of quality assurance for processes, procedures, and conditions in environmental and pharmaceutical laboratories.

Gould's S1 medium a *Pseudomonas*-selective growth medium.

gradostat a multistage chemostat that provides spatial gradients for the study of microbial populations in liquid culture.

graduated cylinder calibrated cylinder used to measure the volume of liquids.

gram (g, gm) unit of mass.

Gram, Christian developer of the Gram reaction/Gram stain method in 1884.

gram-negative characterization of prokaryotic cells that do not retain the crystal violet reagent of the Gram stain when exposed to a decolorization reagent due to the presence of only a thin peptidoglycan layer in the cell wall and therefore stain with the safranin counterstain resulting in a pink to red coloration to the cell when viewed with light microscopy; in contrast to gram-positive.

gram-positive characterization of prokaryotic cells that retain the crystal violet reagent of the Gram stain when exposed to a decolorization reagent due to the peptidoglycan composition of the cell wall and therefore show a purple coloration to the cell wall when viewed with light microscopy; in contrast to gram-negative.

Gram reaction/Gram stain a differential staining procedure used to classify prokaryotic cells in which a series of reagents is applied to an air-dried, heat-fixed suspension of cells on a microscope slide beginning with crystal violet followed by a decolorization step using acetone or alcohol, a water rinse, and a counterstain of safranin.

Graphium fungal genus; member of the hyphomycetes; conidiophores are arranged in a complex structure called a synemma that produces sticky masses of hyaline, ellipsoid or clavate condia; isolated from the air using the Andersen impactor sampler and from wood, plant debris, and soil in cultures; observed microscopically using transparent tape sampling of surface materials; some species may elicit allergic reactions in sensitized individuals; the *Graphium*-state of *Pseudallescheria boydii* is an opportunistic pathogen.

GRAS status generally recognized as safe status.

gravitational sampling collection of bioaerosols using an exposed agar-filled culture plate; does not provide quantitative data and may be only semiqualitative in nature due to interferences in natural settling of particles onto the agar surface; in contrast to active sampling.

gravity plate exposed agar-filled culture plate for collection of bioaerosols using gravitational sampling.

grazing a means of ingestion of masses of microorganisms that are colonized on submerged surfaces that is used by a variety of aquatic organisms such as snails and urchins with scraping mouth organs; in contrast to filter feeding.

green bacteria anoxygenic phototrophic bacteria that conduct photophosphorylation using Bchl *c* and Bchl *d* chlorophyll pigments that are located within chlorosomes, in contrast to the purple bacteria.

green fluorescent protein (GFP) protein produced by some bacteria due to the introduced *gfp* gene that is used as a biosensor for the discrimination of bacteria from background matrix.

green sulfur bacteria autotrophic phototrophic bacteria that oxidize reduced sulfur compounds under anaerobic conditions resulting in the formation of elemental sulfur that is stored outside of the bacterial cell.

gregarious closely scattered over a small area.

groundwater water beneath the earth's surface that occupies the voids in a soil or geologic formation.

groundwater under the direct influence of surface water (GWUDI) a term developed by the United States Environmental Protection Agency to refer to groundwater sources where the environmental, physical, and/or hydraulic conditions are such that pathogens such as protozoan parasites (e.g., *Giardia lamblia*) have either been proven or are likely to travel from nearby surface waters into the groundwater used as a source of potable water; with respect to microbial contaminants, it is regulated as surface water rather than groundwater.

growth an increase in the number of microorganisms.

growth curve depiction of the cycle of a microbial population in culture in which the population increases, stabilizes, and decreases; defined for each microbial population with a lag phase, an exponential phase, a stationary phase, and a death phase.

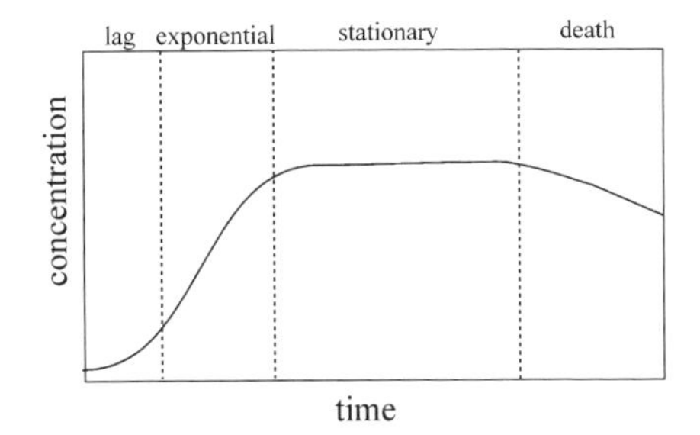

growth limits the range of temperature that defines maximum and minimum growth temperatures for a microorganism.

growth rate (μ) the change in the number of microorganisms per unit time; for bacteria this is expressed as $\mu = \mu_{max} S/K_s + S$, where μ_{max} is the maximum growth rate of the organism growing in a substrate limiting environment, K_s is the saturation constant (the concentration of the growth-limiting substrate allowing growth at half the maximum rate), and S is the actual concentration of the growth-limiting substrate.

GSC gas-solid chromatography; gas chromatography.

GTN glycerol trinitrate; nitroglycerin.

guanine a purine base that is one of the four nucleotides comprising DNA and RNA molecules; forms a bond with cytosine on the opposite strand of the DNA molecule.

guild populations within a community that use the same resources.

guttulate having one or more oily globules.

Gymnomycota a grouping of organisms that produce spores with cell wall structure resembling the fungi but also have vegetative cells that are amoeboid and lack a cell wall resembling the protozoa; slime molds.

H

H̄ symbol for the Shannon-Weaver Index of Diversity.

H_0 null hypothesis.

habitat the physical location where an organism lives.

HACCP hazard analysis critical control point.

hadal zone the vertical area under the bathypelagic zone in the marine environment extending below 6000 meters.

Haeckel, Ernest Heinrich German biologist who proposed a three kingdom classification system for living organisms into animal, plant and protists which included all microorganisms; originated the term microbial ecology in 1866.

half-life time required for a concentration or activity to reach one-half its original concentration; generally used for radiation.

halobacteria generalized term to describe members of the Archaea that are extreme halophiles.

Halobacterium bacterial genus; a member of the Archaea that is characterized as an extreme halophile, gram-negative, chemoorganotrophic organism that is isolated from salted fish, hypersaline lakes; the most studied genus within the halobacteria.

halophile a salt tolerant or salt-requiring organism; microorganisms with specialized enzymes that are in their active configuration only at high salt concentrations.

halorespiring the coupling of reductive dehalogenation and electron transport-coupled phosphorylation catalyzed by specific enzymes in some bacterial species that results in degradation of haloorganic compounds in anoxic polluted soils, aquifers, and sediments; in contrast to use of metal-containing cofactors for reductive dehalogenation by many anaerobic bacteria.

hanging drop mount suspension of a single drop of sample in liquid over the concave surface of a special microscope slide for viewing motility.

HAPs hazardous air pollutants.

Harris slide culture method a culture technique for the identification of fungi in which the sides of a small section (~2 cm × 2 cm) of agar medium are cut from one area of a agar-filled petri plate, placed on the agar of another area of the plate and then inoculated with the test organism and covered with a sterile glass coverslip so that following incubation the coverslip can be removed, stained, and viewed without disruption of the delicate structures; variation of the Riddell slide culture method.

harvest in cell culture, describes the collecting of cells or the growth medium in which cells exposed to a sample were grown, so that the cells or medium can be analyzed or the cells can be used as a seed for further propagation of the cell culture.

hathi-gray elephant gray in color.

haustorium projection of fungal hyphae into host cell cytoplasm.

Hazard Analysis Critical Control Point (HACCP) a U.S. Food and Drug Administration proposed regulation for food processing facilities that is designed to decrease the concentration of target pathogens in products.

hazard identification the area of risk assessment that encompasses the determination of whether a particular agent is or is not associated with a particular health effect.

hazardous air pollutants (HAPs) airborne chemicals that cause serious health and environmental effects.

hazardous waste/hazardous material general term for any waste materials that, when released to the environment, may pose a health hazard to exposed humans or animals, or to the quality of the environment.

haz mat hazardous material.

healthy worker effect the observance that workers exhibit fewer symptoms than the general population because generally the aged and disabled are excluded from the workforce.

heating, ventilation and air conditioning system (HVAC) the mechanical system designed to control the temperature and humidity of the indoor environment and distribute conditioned air throughout a building.

heat-shock protein (HSP) intracellular protein that increases in concentration during stress.

hedonic tone a judgment of the relative pleasantness or unpleasantness of an odor.

HeLa cells human epithelial cells from a cervical carcinoma (from a woman named Henrietta Lacks) that have been adapted to grow *in vitro* as a continuous cell line; used for the detection and cultivation of viruses.

helical in the shape of a coil or spiral; many viruses have capsids that exhibit helical symmetry in which the capsomeres are packed around the nucleic acid in the shape of a spiral.

helicoid coil-like in appearance.

helix spiral structure containing repeating sequences.

Helminthosporium fungal genus; isolated from plants, some species are phytopathogens; colony on malt extract agar is downy to wooly in texture and olive brown to black in color; brown, septate hyphae and stiffly erect, determinate conidiophores; obclavate poroconidia that are transversely multiseptate; in contrast to *Bipolaris*, *Dreshlera* and *Exserohilum*.

hematopoietic pertaining to or affecting the formation of blood cells.

hematuria presence of blood or blood cells in the urine.

hemocytometer device used for the enumeration of microorganisms in liquid suspension using light microscopy; grid pattern permits the calculation of the number of organisms per ml of liquid.

hemolysin toxin capable of causing the lysis of red blood cells.

HEPA filter high efficiency particle arresting filter.

hepatitis inflammation of the liver.

hepatitis A virus a small (27–32 nm), nonenveloped single-stranded RNA virus, icosahedral in shape; member of the Picornaviridae family; primary site of infection is the liver; transmitted by the fecal-oral route of transmission, it causes water- and food-borne illness; very resistant to adverse environmental conditions.

hepatitis virus a general term used to describe one of a number of unrelated viruses that infect the liver, resulting in jaundice; to date, hepatitis viruses A through G have been described.

hepatotoxin a compound that is toxic to the liver.

HEPES 4-(2-hydroxyethyl)-1-piperazine-ethanesulfonic acid; a buffer commonly used in cell culture applications.

Hesse, Walter initiated studies of airborne microorganisms in 1882 and first reported on the use of agar as a solidifying agent for culture media as suggested by his wife, Fannie.

heterocyst a specialized cyanobacterial cell that conducts nitrogen fixation.

heterofermentation the fermentation of a sugar to a mixture of products.

heterothallic sexual reproduction of a fungus that requires mating of cells from two sexually compatible strains; in contrast to homothallic.

heterotroph/heterotrophic a microorganism that utilizes organic carbon as a nutrient source for growth.

heterotrophic plate count (HPC) procedure used for the enumeration of culturable heterotrophic bacteria in water; formerly termed standard plate count.

heterotrophic potential the ability of microbial populations to utilize an organic substrate.

heterotrophic succession shift in microbial populations due to the insufficient supply of organic material for the survival of heterotrophic microorganisms.

Hfr high frequency of recombination.

hierarchical stepwise regression statistical analysis in which a single variable or clusters of variables are examined by stepwise regression but the order of the introduced variables is predetermined.

high efficiency particle arresting (HEPA) filter a filter composed of randomly positioned glass fibers that is highly efficient (>99.9%) at removing submicron (down to approximately 0.3 microns) particles from air.

high frequency of recombination (Hfr) a bacterial cell in which the F plasmid has been transferred from an F+ cell to an F– cell; in contrast to F+ cells, the F plasmid is incorporated into the bacterial chromosome and replicates with it, thus progeny are also Hfr.

high performance liquid chromatography (HPLC) chromatographic separation technique used for qualitative and quantitative analysis of large nonvolatile molecules; the sample is introduced into a liquid mobile phase that flows over a solid or chemically bonded stationary phase and detected primarily using ultraviolet light, fluorescence, or mass spectrometry.

high throughput screening (HTS) methods used in pharmaceutical facilities to rapidly test fermentation extracts for antibiotic activity.

high volume small surface sampler (HVS3) vacuum mounted cyclone forced air flow sampler originally designed as a collection system for determination of asbestos.

hilar appendage an attachment between a basidiospore and the sterigma.

hilum a mark or scar; generally used to describe the area of a fungal spore where it was attached to the conidiogenous cell.

hirsute hairy in appearance.

histogram presentation of data using the length of a bar to denote value; similar to a bar chart plot except that there are no spaces between the bars.

Histoplasma capsulatum fungal species; isolated from nitrogen-rich soils contaminated by bats and birds; endemic in the Mississippi and Ohio River valleys of the United States and encountered in other subtropical and tropical regions and temperate river basins; when isolated at 25°C the colony on malt extract agar is wooly to granular in texture that is white in color becoming brownish on the surface and yellowish on the reverse producing macroconidia that are thick-walled, hyaline, and unicellular with either a smooth or warty appearance to the walls and microconidia that are hyaline and unicellular with smooth or rough walls; at 37°C the colony is creamy in texture and cream in color; hyaline hyphae are septate producing budding yeast cells.

histoplasmosis respiratory disease caused by inhalation of spores of *Histoplasma capsulatum*

hockey stick spreader glass rod bent to a 90° angle for use in distributing a liquid sample evenly across the surface of an agar-filled petri plate.

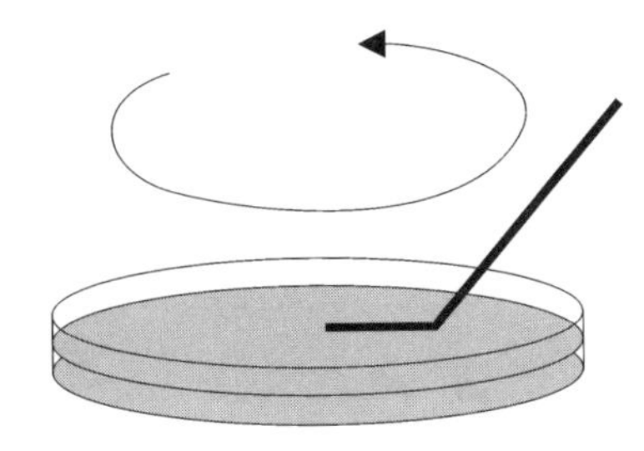

holdfast special fungal hyphae used for attachment.

hollow-stem augering drilling technique used to obtain samples in shallow, unconsolidated subsurface environments that does not require the use of drilling fluids that often result in contamination of the samples.

homeostasis a compensating mechanism that acts to maintain steady state conditions; the capacity of a community to react to biotic and abiotic conditions and maintain stability and integrity.

homoacetogen a bacterium that produces acetate from H_2 and CO_2 or as the result of the fermentation of sugar.

homofermentation the fermentation of sugar with lactic acid as the only product.

homothallic spontaneous sexual reproduction of a single fungal isolate by identical nuclei undergoing fusion; in contrast to heterothallic.

Hooke, Robert an English philosopher and experimenter who described his microscopic observations of fungi and protozoa in 1665.

horizontal gene transfer the transfer of genetic information between organisms in the same niche; also termed lateral gene transfer; in contrast to vertical gene transfer to progeny.

Hormodendrom fungal genus; former terminology for *Cladosporium*.

host an organism that is infected by another organism such as a parasite or virus; the host may be harmed as a result of the infection.

host range the breadth of organisms that can be infected by a particular pathogen.

hot plate device designed to heat liquids in a beaker or flask at various temperature settings, often used with a magnetic stir bar to provide mixing.

hot start PCR a variation of the traditional polymerase chain reaction amplification method in which the reaction mixture is separated from the *Taq* polymerase by a wax layer, which does not melt until the temperature exceeds that required for DNA denaturation, thereby preventing nonspecific hybridization that can occur at lower temperatures.

hot vent a hydrothermal vent that emits mineral-rich water of 270-380°C and forms a black cloud of precipitate; also termed "black smoker."

Howard mold chamber device for the enumeration of fungal spores in liquid suspension using light microscopy.

HP hypersensitivity pneumonitis.

HPC heterotrophic plate count.

HPLC high performance liquid chromatography.

HSP heat-shock protein.

HTS high throughput screening.

hülle cells refractile, thick-walled cells characteristic of some *Aspergillus* spp.

human-to-human transmission transfer of infectious microorganisms from one person to another without a vector, vehicle or fomite intermediary.

humic acid organic polymers found in soils that are formed during the degradation of lignin that are resistant to biodegradation.

Humicola fungal genus; light gray colony on malt extract agar with somewhat darker reverse; solitary brown globose to elongated smooth-walled conidia that are surrounded with a slimy sheath and small, hyaline obovoid conidia; similar in structure to *Nigrospora*; isolated from soils and plant debris and in products with high cellulose content.

humidification/humidifier increase in the relative humidity of the air by the aerosolization of water vapor.

humidity cell test a 20-week assay standardized by ASTM International; designed to model the natural weathering of waste material with cycling of wet and dry air on specific days followed by collection of a water leachate for analysis for physical parameters (e.g., pH, acidity, alkalinity) and dissolved metals.

humus stable organic matter that accumulates in soils and sediments that is considered to be inert, but serves as a terminal electron acceptor and has an active role in anaerobic oxidation of some organic substrates.

Hungate technique methodology for the isolation of obligate anaerobes in which the preparation of media and culture of organisms is conducted under strict anaerobic conditions using oxygen-free gases and air-tight vessels or glove-box enclosures.

HVAC heating, ventilation and air conditioning system.

HVFS an adaptor for the high volume small surface sampler.

HVS3 high volume small surface sampler.

hyaline colorless, nonpigmented.

hybrid the progeny of genetically different parents (e.g., from different species).

Hybrid Single-Particle Lagrangian Integrated Trajectory (HYSPLIT) model computer program designed to calculate the trajectory of airborne fungal spores of agricultural importance.

hybridization the construction of a double-stranded nucleic acid by complementary base pairing of two single-stranded nucleic acids in the laboratory.

hydro extraction a dialysis process using polyethylene glycol to concentrate viruses from a water sample.

hydrogen cycle the transformation of hydrogen in reservoirs by biological activity; anaerobic fermentation and photosynthesis produce free H_2 gas that is metabolized anaerobically to NH_3, hydrogen sulfide, or methane while oxidative metabolism in aerobic environments by chemolithotrophs produces water.

hydrolysis the reaction of water with a salt resulting in the formation of an acid or a base.

hydrolytic promotes hydrolysis.

hydrophilic molecule that interacts readily with water due to the presence of polar groups; contrast with hydrophobic.

hydrophobic molecule that does not interact readily with water, e.g., lipid; contrast with hydrophilic.

hydrosphere the region of the earth covered with water in the form of liquid water, ice, or water vapor, including groundwater, surface water, and atmospheric moisture.

hydrostatic pressure the force exerted by the weight of a water column that increases approximately 1 atmosphere for every 10 meter of depth and will be even greater in saline waters than freshwater due to increased weight of saline water; in contrast to atmospheric pressure.

hydrothermal vent an underwater hot spring.

hydroxy fatty acids frequently occurring in gram-negative bacteria as components of lipopolysaccharides or poly-β-hydroxybutyrate and found in some actinomycetes.

hygrometer instrument for the measurement of relative humidity.

hygroscopic readily absorbing and retaining moisture.

hyperplasia excessive growth, abnormal increase in the number of normal cells in a tissue; symptom of some plant diseases caused by microorganisms.

hypersensitivity an immunological response to a foreign substance caused by a cellular immune response or an antigen-antibody reaction.

hypersensitivity pneumonitis (HP) cell-mediated immune reaction; may be the result of Type III allergies and Type IV delayed hypersensitivity; formation of granulomas in the lung alveoli and pulmonary fibrosis resulting from an immunologic response following inhalation route of exposure to thermophilic actinomycetes, some fungi, chemicals, and protein of some animals and birds.

hyperthermophile/hyperthermophilic descriptive of the temperature requirements of microorganisms with an optimal growth temperature greater than 80°C.

hypertonic having a higher osmotic pressure than the surrounding liquid; contrast with hypotonic.

hypha singular term of hyphae.

hyphae septate or nonseptate filaments of fungi.

hyphal fragment a fractioned segment of hyphae.

hyphomycete classification of fungi in the Deuteromycetes with conidiophores that are formed on simple or aggregated hyphae producing conidia; members of this group are often referred to as molds.

hypogeous fungi subterranean fungi that produce sporocarps; e.g., truffles.

hypolimnion thermal region of water below the thermocline that is characterized by low temperatures and low oxygen concentrations with the poor light penetration restricting photosynthesis and respiration depleting the existing oxygen, but mineral nutrients are relatively abundant.

hypoplasia stunted growth; symptom of some plant diseases caused by microorganisms.

hypotonic having a lower osmotic pressure than the surrounding liquid; contrast with hypertonic.

I

IARC International Agency for Research on Cancer of the World Health Organization.

ice bath suspension of ice in water or dry ice in ethanol for rapid cooling of material.

ICNV International Committee on Nomenclature of Viruses.

ICP-AES inductively coupled plasma atomic emission spectroscopy.

ICP-MS inductively coupled plasma mass spectroscopy.

icosahedral of or pertaining to an icosahedron.

icosahedron an object whose surface is composed of twenty equilateral triangular facets and twelve vertices.

ICTV International Committee on the Taxonomy of Viruses.

ID_{50} infectious dose 50.

IDHLs immediate danger to life or health exposure levels.

idiopathic pulmonary hemorrhageidiopathic pulmonary hemosiderosis bleeding in the lung experienced by some infants resulting in death; possible association with *Stachybotrys*-contaminated indoor environments and increased risk with exposure to environmental tobacco smoke.

Imhoff tank a septic tank design in which anaerobic conditions are more strictly maintained than traditional systems.

immaculate pure, unspotted.

immediate danger to life or health exposure levels (IDHLs) exposure levels established by the National Institute for Occupational Safety and Health in the 1970s for 398 chemicals.

immigration the movement of a microbial population into an environment; in contrast to emigration.

immiscible solutions that do not mix; in contrast to miscible.

immunoassay the use of antigen-antibody reactions to detect the presence of a microorganism or immune response.

immunocompetent able to recognize and react to a foreign substance in the body.

immunocompromised failure of the immune system to elicit a response; often the result of an immunodeficiency disorder such as AIDS, exposure to radiation, or immunosuppressive drugs; this may render the individual more susceptible to infection by certain pathogens and may result in a more severe infection than experienced by nonimmunocompromised individuals.

immunofluorescence assay the use of a light emitting dye in an immunoassay; the dye is bound to an antigen for the detection of the corresponding antibody, or to an antibody for detection of antigen in the sample.

immunogenimmunogenic an antigen that can elicit an immune response.

immunoglobulin (Ig) an antibody.

immunomagnetic separation (IMS) a technique used in molecular biology in which paramagnetic beads coated with antibodies specific for the target of interest are used to separate those target organisms from a liquid sample upon application of a magnet.

immunosuppressed having a lower resistance to disease than the typical individual, may be caused by exposure to chemotherapeutic drugs.

immunotoxic negatively affects the immune system.

impaction sampler a forced air flow sampler designed for the collection of bioaerosols that utilizes the inertia of the particles to force their deposition onto a solid or semisolid surface; collection is dependent on the properties of the airborne particle (e.g., size, shape, density, velocity) and the physical properties of the sampler (e.g., dimensions of the inlet nozzle, air flow pathway); solid surface devices use tape or a sticky film coating on the surface of a microscope slide, or coverslip for collection of airborne fungal spores and pollen with the samples analyzed using light microscopy; semisolid surface devices use an agar-filled petri plate or individual agar-filled well as the collection surface for the monitoring of airborne bacteria or fungal spores with culture-based analysis.

imperfect state the asexual form of a fungus.

impingement sampler a forced air flow sampler designed for the collection of bioaerosols into a liquid collection medium; liquid collection permits a variety of analysis methods including culture, microscopy, biochemical assay, and molecular biology techniques; sample can be diluted or concentrated prior to analysis to avoid problems with upper limits of quantitation or lower limits of quantitation encountered with impaction samplers.

implementing procedure (IP) protocol.

IMS immunomagnetic separation.

inactivate/inactivation the disruption or prevention of an activity; often refers to causing the loss of infectivity of an organism.

incidence the number of new occurrences of an event during a specified period of time; in contrast to prevalence.

incidence density the number of occurrences within a defined area.

incidence rate the number of new occurrences of an event (e.g., illness) per population at risk within a defined period of time.

inclusion bodies granules or chemical units within a cell separated from the cytoplasm by a lipid membrane.

incompatability the inability of two or more materials to occupy the same location or be present in a mixture.

incompatability groups plasmids that are not able to co-replicate in a single bacterium.

incubation period the time between exposure and the amplification of a microorganism in a host to a level where signs and/or symptoms of the infection are manifest.

incubator an enclosed, temperature-regulated chamber for the growth of microorganisms in culture.

indehiscent not splitting; in contrast to dehiscence.

independent variable that condition which is being changed or manipulated during the conduct of an experiment.

indeterminate unlimited; a conidiophore or conidiogenous cell that continues to grow after the first conidium is formed; in contrast to determinate.

India ink a dye used as a negative stain, but care must be used as this material often is contaminated with bacteria.

indicator organism a surrogate organism that is used to demonstrate a condition or the potential presence of another organism which is difficult to detect; the indicator should be at least as resistant or persistent, be present in greater concentrations, and should be readily detectable; the presence of *Escherichia coli* serves as an indicator of fecal contamination in water and food samples.

indigenous naturally occurring; also termed autochthonous.

indirect causation a result is not associated with a single factor but requires additional factors or intermediate steps; in contrast to direct causation.

indirect fluorescent antibody staining technique in which a nonfluorescent antibody is directed to the antigen of interest and then a fluorescent antibody is applied that is directed to the nonfluorescent antibody; in contrast to direct fluorescent antibody staining.

indirect transmission transfer of a pathogen to a new host via a vector, fomite, or a vehicle.

infect to enter another organism and multiply within it.

infection focus descriptive of the area of a crop with an infectious disease.

infection thread a cellulose tube though which *Rhizobium* cells travel to infect root cells.

infectious capable of invading a host cell and multiplying thereby causing an infection.

infectious dose the number of organisms required to produce an infection in an exposed individual.

infectious dose 50 the infectious dose required to produce an infection in 50% of the test population.

inferential statistics used to determine the probability that a conclusion based on analysis of data from a sample is true and can be used to generalize from a set of sample data to a larger population.

inflammation, inflammatory response swelling, heat, pain, and redness associated with a host response to exposure to a foreign substance.

infrared spectrometry (IR) an analytical technique used for the characterization of surfaces and interfaces between surfaces.

infundibuliform funnel-shaped.

ingestion route of exposure microorganism or microbial by-product enters the body by eating contaminated food or water; contaminant generally must pass through the digestive process in the stomach before absorption occurs in the intestine.

inhalation route of exposure microorganism or microbial by-product enters the body by breathing contaminated air; the exposure dose needed to initiate an adverse reaction may be one-tenth that required via the ingestion route of exposure because sensitive lung tissue is directly exposed to the contaminant without passage through the digestive system.

inhibition the prevention of a function; this condition maybe temporary.

inoculate to add microbial cells to a growth medium or viruses to a host.

inoculating loop a flat, open circle device used to apply microorganisms to a surface or liquid.

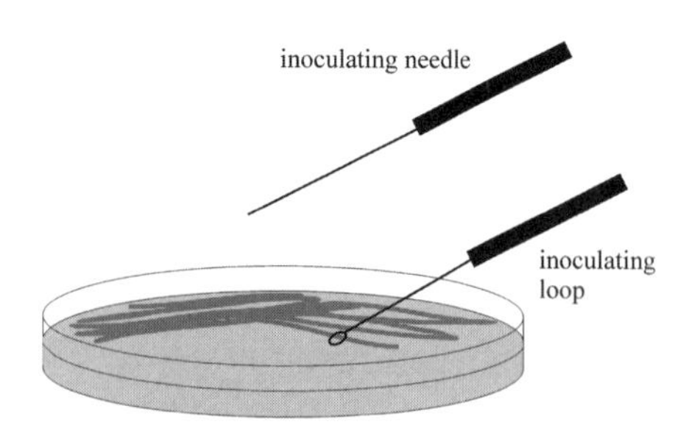

inoculating needle thin wire generally used for pinpoint application of microorganisms to a surface.

inoculative augmentation the introduction of a microbial pathogen into the environment with the expectation that it will cycle through the host population and provide an effective control of the host.

inoculum the microorganisms, cells, or other biological material that are added to growth medium to start a culture.

inoperculate without a lid; in contrast to operculate.

inorganic compound chemical that does not contain carbon.

Inoviridae a family of filamentous or rod-shaped single-stranded DNA bacteriophage.

inovirus single-stranded filamentous DNA bacteriophage in the family Inoviridae.

insertion the placement of a piece of DNA within the sequence of a gene.

in situ Latin term for the original place; used to denote experiments conducted at the site or on location.

institutional review evaluation by the administrative body overseeing conduct of research.

INT 2-(*p*-iodophenyl)-3-(*p*-nitrophenyl-5-phenyl tetrazolium chloride; a dye used to determine if microorganisms in a sample are respiring.

integration the incorporation of DNA into another genome.

integument an outer coating.

interaction a collaborative action between two or more entities.

intercalary between the apex and base.

interference contrast microscopy technique in which two beams of light are optically produced, one carrying the image and the other serving as a reference beam that causes interference; used for the resolution of fine subcellular detail.

interlaboratory quality control a program of requirements and laboratory practices among a group of participating laboratories that conduct the same analysis.

internal transcribed spacer (ITS1 and ITS2) a DNA sequence that separates specific repeated rRNA genes (e.g., the 16S and 23S rRNA gene or the 23S and 5S rRNA gene); used for identification of closely related prokaryotic organisms at the species or even strain level.

International Code of Nomenclature of Bacteria publication describing the rules for naming of bacteria.

International Committee on Nomenclature of Viruses (ICNV) determining body for the naming of viruses that established in 1966 that virus classification would not be determined by the host that is infected, but by the presence of DNA or RNA and the symmetry and other characteristics of the virion.

International Committee on the Taxonomy of Viruses (ICTV) name for the International Committee on Nomenclature of Viruses (ICNV) beginning in 1973; operates under the auspices of the Virology Division of the International Union of Microbiological Societies.

interquartile range the middle 50% of data; also termed mid-spread.

interstitial situated in the space between things.

intertidal zone region of the ocean at the seashore with alternating periods of flooding and drying at high and low tides; the littoral zone in the marine environment; represents the area between the marine ecosphere and the lithosphere.

interval data/interval variables data that exist in an ordered category where the differences between values are equal and the zero point is arbitrary; in contrast to a ratio variable.

interval width the range of values that are represented in an interval when arranging interval data, generally the width is considered in multiples of 2, 5,

10, 20, 100 such as 1–3 or 5–15 with the greater widths having more detail that is lost in grouping of the data.

intine inner wall of a fungal spore.

intracellular inside of a cell.

intramatrical within a substrate; in contrast to extramatrical.

intron a noncoding sequence of nucleotides within a gene that is transcribed into pre-messenger RNA but then cut out of the molecule and subsequently degraded.

inverted microscope a microscope with the light source above the stage and the ocular and objectives placed underneath the stage; this arrangement permits the placement of the specimen over the objective and is used to view cells in tissue culture flasks.

in vitro Latin term "in glass" used to denote experiments conducted outside of a cell.

in vivo Latin term "in cell" used to denote experiments conducted in living cells.

iodine used as a mordant in the Gram reaction; used to disinfect surface water.

ionizing radiation the interaction of radiation with matter that produces unstable ions and free radicals that interact with living organisms in a destructive way.

IP implementing procedure.

IPU phenylurea herbicide isoproturon (3-(4-isopropylphenyl)-1,1-dimethylurea) used for pre- and postemergence control of annual grasses and broad leaf weeds in crops.

IR infrared spectrometry.

irgalan black dye used in fluorescent staining methods to provide a black background and minimize autofluorescence of the filter.

iron cycle the transformation of iron in reservoirs by biological activity; the oxidation of ferrous iron (Fe^{2+}) to ferric iron (Fe^{3+}) under aerobic conditions and the reduction of Fe^{3+} to Fe^{2+} under anaerobic conditions.

irritant asthma asthma resulting from high exposure to particulate material.

isabellin pinkish-cinnamon in color.

isoelectric point the pH at which a protein has a net electrical charge of zero; at pH levels below the isoelectric point, the protein has a net positive charge due to the protonated state of ionizable groups on the protein molecule; at pH levels above the isoelectric point, the protein has a net negative charge.

isolation the separation of two or more entities.

isolation streak method to separate organisms present in a liquid sample so that each colony grows as a discrete unit on the surface of an agar-filled plate; microbial suspension is streaked onto the agar surface with an inoculating loop in quadrants resulting in decreasing concentrations; often the loop is flame sterilized between quadrants.

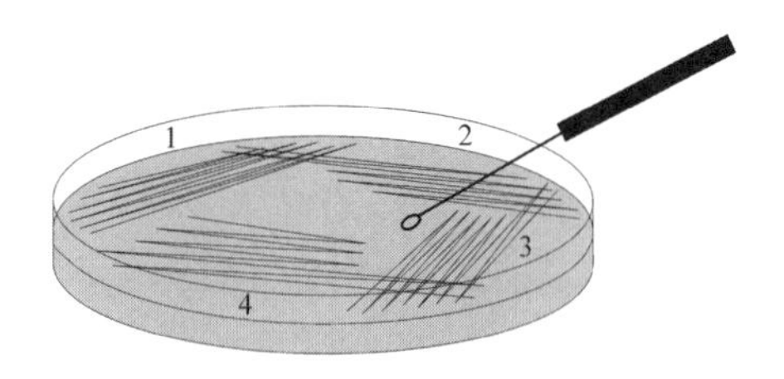

ITS1 and ITS2 internal transcribed spacers.

J

J symbol for equitability.

Japan Collection of Microorganisms (JCM) culture collection entity located in Wako, Japan that catalogs and sells standard strains of microorganisms.

jaundice a yellowing of the skin, mucous membranes, and eyes due to rapid destruction of blood cells or abnormalities of the liver; a symptom of hepatitis.

JCM Japan Collection of Microorganisms.

joist horizontal beam as a structural element of a building.

joule (J) unit of energy equal to 10^7 ergs.

K

kappa statistic proposed by J. Cohen in 1960 as a means to determine if the agreement in evaluations involving subjective analysis recorded by two independent observers is greater than what would have occurred by chance.

kb kilobase.

Kelly's medium culture medium for the isolation of *Borrelia* spp.

ketocaproylhomoserine lactone/*N*-β-ketocaproylhomoserine lactone an autoinducer found in *Vibrio fischeri* that induces an enzyme when it reaches a critical level in the culture medium during growth resulting in luminescence.

kg kilogram.

kilobase (kb) a 1000-base segment of nucleic acid.

kilobase pair a segment containing 1000 base pairs.

kilogram (kg) a unit of weight equivalent to 1000 grams.

kingdom the second highest taxonomic ranking; the taxonomic ranking below domain.

Klebsiella bacterial genus, member of the family Enterobacteriaceae; nonmotile, gram-negative bacilli some of which are human pathogens while others are commensals in humans and animals, or phytopathogens.

Klebsiella oxytoca bacterial species; opportunistic pathogen with potential environmental exposure via aerosols generated during wastewater treatment practices.

Klebsiella pneumoniae bacterial species; human pathogen with potential exposure via aerosols generated during wastewater treatment practices.

Kluyver, A.J. (1888–1956) microbial physiologist who studied oxidative, fermentative, and chemoautotrophic microorganisms, and believed that within the diverse microbial population there were unifying metabolic features.

Koch, Robert (1843–1910) developer of solidified media for the pure culture of microorganisms; discoverer of *Bacillus anthracis* as the causative agent of anthrax (1877), *Mycobacterium tuberculosis* as the causative agent of tuberculosis (1882), and *Vibrio cholerae* as the causative agent of cholera (1883).

Koch's postulates criteria proposed by Robert Koch to establish the relationship of a disease-causing organism with the resulting disease that include the presence of the organism in every incidence of the disease, the absence of the organism when the disease was not present, and the isolation of the organism, inoculation into a healthy individual and subsequent, resulting disease.

Kogure technique the use of nalidixic acid and yeast extract with acridine orange and light microscopy to enumerate swollen or elongated cells as viable.

Korarchaeota a kingdom of hyperthermophilic Archaea.

Kreb's cycle citric acid cycle.

K strategists microorganisms that depend on physiological adaptations to environmental resources or the carrying capacity of the environment for continued survival within a community so they reproduce slowly and are successful in environments that are limited in nutritional resources.

Kuntzing, Friedrich concluded independently of T. Schwann, but in the same year (1837), that alcoholic fermentations were caused by yeasts.

kurtosis referring to the nature of a plot of data as curved or flat when represented in a graph.

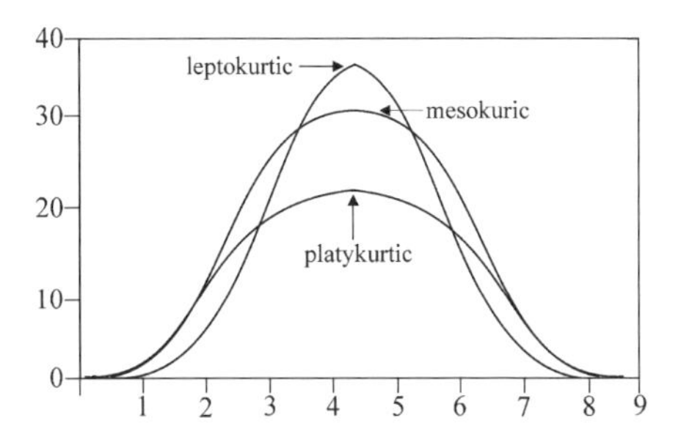

Kurung-Yegian agar a fungal culture medium used for the conversion of *Histoplasma capsulatum* and *Blastomyces dermatitidis* to the yeast phase.

L

LAB lactic acid bacteria.

laccases multi-copper blue oxidase enzymes implicated in fungal conidia production, lignin degradation, pathogenicity, and fruiting-body production.

lactic acid bacteria (LAB) general description of gram-positive bacteria that produce lactic acid as the major or sole product of fermentation; used in the production of a wide range of fermented foods, organic acids, and antibiotics; some produce bacteriocins.

Lactobacillus bacterial genus; gram-positive, anaerobic or facultatively anaerobic bacilli; many species important in the production of cheese, yogurt, and sauerkraut; several species have demonstrated antifungal activity and can serve in biopreservation; other species are proposed as protobiotics.

Lactobacillus plantarum bacterial species; lactic acid bacterium used to initiate fermentation in food products.

lactococcal phages bacteriophages that infect *Lactococcus* species; of the 12 species, three are virulent types that infect *Lactococcus lactis* causing fermentation failures in the dairy industry that result in slow acid formation and inferior products.

Lactococcus lactis bacterial species; active in the mesophilic fermentation of dairy products including the production of cheddar cheese, sour cream, and buttermilk; serves as a vehicle for mucosal vaccines as it can deliver antigen to mucosal immune system without colonizing the host tissue.

lactophenol cotton blue a mounting and staining reagent used for the examination of fungal preparations with light microscopy.

lactose broth a liquid nutrient medium containing lactose sugar that is used for the detection of total coliform bacteria or fecal coliform bacteria.

lageniform flask-shaped.

lag phase initial period of time in the growth curve of microorganisms in which growth does not occur immediately; the period to time prior to exponential growth.

lake snow macroscopic organic aggregates containing algae, zooplankton molts, organic and inorganic debris, and bacteria that are involved in recycling of nutrients in lakes.

LAL assay *Limulus* amebocyte lysate assay.

lamellae the flat photosynthetic membrane system found in some purple bacteria.

laminar flow the flow of air along a defined path; used in biological safety cabinets to minimize exposure of laboratory technicians to material within the cabinet.

lanceolate lance-shaped.

landfill acreage used for the disposal of organic and inorganic waste designated as a sanitary landfill when refuse is covered daily with a layer of soil.

lanose woolly in appearance.

lantibiotic a class I modified bacteriocin.

LAS linear alkylbenzene sulfonate.

latent infection an infection that is asymptomatic, but may become symptomatic under the proper conditions.

latent virus a virus whose genome is integrated into the host's genome, but is not expressed; upon activation (e.g., by stress or exposure to ultraviolet irradiation), infective virus particles are produced and symptoms of infection appear.

lateral gene transfer horizontal gene transfer.

lawn bacterial lawn.

LDL lower detection limit/lower limit of detection.

leach field an area of land to which wastewater is applied, as the wastewater percolates through the soil, removal of some contaminants occurs; used with a septic tank.

leaching the removal of metal from ore by chemical or microbial activity; the transport of dissolved materials from upper soil layers deeper into the subsurface.

least-squares analysis statistical analysis of data that creates a regression line indicating the least square sum between the actual data and fitted data.

LEB *Listeria* enrichment broth.

lectins multivalent carbohydrate-binding proteins produced by bacteria, plants, and animals; involved in the binding of bacterial fimbriae to surfaces with sugar residues; involved in the association of *Rhizobium* spp. to host plant root hairs.

Leeuwenhoek, Antonie van (1632–1723) developer of simple microscopes in the 17th century who described in detail his observations of bacteria in natural habitats and environmental influences on microorganisms.

Legionellaceae bacterial family; characterized as intracellular parasites or endosymbionts of free-living parasites.

Legionella bacterial genus; aerobic, motile, fastidious, gram-negative bacilli that multiply as intracellular parasites of humans and protozoa; 42 species comprising 64 serogroups; commonly isolated from freshwater environments with 40% of fresh waters positive by culture and 80% positive by polymerase chain reaction amplification; protozoa are the natural host with multiplication reported in 13 species of amoebae and 2 species of ciliated protozoa; growth in the laboratory forming white colonies 2–4 mm in diameter with a "ground glass" appearance ringed with blue, purple, red or green iridescence at the edge of each colony at 35°C in a humidified 2.5% CO_2 atmosphere on buffered yeast charcoal extract medium containing L-cysteine; additional amendments may be added as selective agents depending on species; poor survivability in dry environments; requires temperatures above 25°C to multiply but are killed between 44–55°C; infrequently cause human disease.

***Legionella*-like amoebal pathogens (LLAPs)** bacteria that are currently classified in the genus *Sarcobium*, but are genetically closely related to the genus *Legionella*; capable of growing intracellularly, and can be detected using immunoassay and polymerase chain reaction amplification methods designed for *Legionella*.

Legionella pneumophila bacterial species; causative agent of approximately 90% of documented legionellosis cases; 15 serogroups with 82% of all legionellosis cases caused by serogroup 1 (Lp1).

***Legionella pneumophila* serogroup 1 (Lp1)** currently 10 subtypes recognized using monoclonal antibody detection methods.

legionellosis human respiratory disease caused by inhalation exposure to *Legionella* characterized by symptoms recognized as Legionnaires'disease or Pontiac fever; primarily sporadic cases with approximately 4% of the 15,000 annual cases in the United States classified as outbreaks.

Legionnaires' disease progressive pneumonic disease first reported following the 1976 American Legion convention in Philadelphia resulting in 239 cases and 24 deaths; caused by inhalation exposure of a susceptible human host to a sufficient quantity of select serogroups of infectious *Legionella pneumophila*; 2–10 day incubation period with a <5% attack rate; host risk factors include cigarette smoking, diabetes mellitus, cancer, end-stage renal disease, and acquired immune deficiency syndrome; no documented cases of human-to-human transmission.

length-based sampling method of selecting individuals in an epidemiology survey based on screening for disease over an extended period of time.

lentic stagnant, nonflowing water.

lenticular shaped like a lens that is convex on both sides.

LEPCs Local Emergency Planning Coordinators.

leptokurtic reference to the kurtosis of a graph when it is more peaked; in contrast to platykurtic.

Leptothrix bacterial genus; sheathed bacteria with iron or manganese oxides encrusting the sheath.

LeTx anthrax lethal toxin.

Leuconostoc mesenteroides bacterial species; phytopathogen; causative agent of postharvest decay of tomatoes.

Leucosporidium fungal genus; basidiomycetous yeast found in marine environments.

Leviviridae a family of single-stranded RNA, icosahedral bacteriophage; MS-2 is the type species.

L-form/L-phase variant bacteria that lack a cell wall but maintain the presence of peptidoglycan; in contrast to Mycoplasmataceae.

lichen mutualistic relationship between certain algae or cyanobacteria and fungi.

Liebig, Justus (1803–1873) German chemist who believed that yeasts and other microorganisms were products of fermentation rather than causative agents, and developer of Liebig's Law of the Minimum.

Liebig's Law of the Minimum theory that the biomass of any organism will be determined by the nutrient present in the lowest concentration in relationship to the requirements of that organism.

ligases enzymes that catalyze the linking together of two molecules.

light microscopy microscopic method using visible light to visualize intact cells; specimens are viewed as wet mounts, transparent tape samples, tape lifts, or heat-fixed and stained preparations; in contrast to atomic force microscopy; scanning electron microscopy; and transmission electron microscopy.

lignin a three-dimensional plant polymer, characterized by a high concentration of aromatic rings, that is relatively resistant to biodegradation although it is degraded by white rot fungi, some bacteria, and other fungi; in wood it is associated with cellulose and hemicelluloses to provide strength.

limnetic zone region of open water away from the shore that penetrates to the compensation depth and where primary production is predominated by planktonic algae.

***Limulus* amebocyte lysate (LAL) assay** an *in vitro* bioassay used to detect endotoxin and also has been used to detect β-(1 → 3)-D-glucan.

Linnaeus, Carl Swedish botanist who established the first classification system for microorganisms consisting of a single genus, *Chaos*, in 1759.

lipid organic molecules (e.g., fatty acids, waxes, and steroids) that are not soluble in water.

lipid A the least variable of the three regions of the lipopolysaccharide in the outer membrane of gram-negative bacteria; responsible for toxigenicity of endotoxins.

lipoglycan the lipopolysaccharide in *Mycoplasma* and *Thermoplasma*.

lipophilic attracted to oils and fats.

lipopolysaccharide (LPS) component of the gram-negative bacterial cell wall that is composed of a complex of sugars and fatty acids.

liquid nitrogen storage the archival of small quantities of reagents or samples in nitrogen gas that is cooled to <–77°C.

Listeria bacterial genus; bluish gray colonies of regular, short, gram-positive, aerobic or facultatively anaerobic bacilli that are motile by a few peritrichous flagella; widely distributed in water, mud, sewage and vegetation, and excreted in the feces of humans and animals; some species pathogenic to humans.

Listeria monocytogenes bacterial species; ubiquitous in the environment and raw foods; tolerant of low temperatures and high salt concentrations; foodborne pathogenic organism associated with contamination of processed food products, especially lightly salted and chilled foods with a prolonged shelf life that are consumed without additional heating.

liters/min (LPM) rate measurement of the volume of air or liquid per unit time; 1 cubic foot per minute = 28 liters/min = 0.028 m^3/min.

lithobiotic rock-inhabiting; microorganisms that colonize rock surfaces or the rock fabric adjacent to the surface; in contrast to endolithic.

lithosphere the outer layer of the earth consisting of the crust and upper mantle.

littoral zone region of lake water in which light penetrates to the bottom; generally shallow in depth and dominated by submerged higher plants and attached epiphytic algae; the intertidal zone in marine environments.

Ljungdahl-Wood pathway acetyl-CoA pathway.

LLAPs *Legionella*-like amoebal pathogens.

lobulate having lobes.

logistic regression statistical analysis used when the dependent variable is dichotomous and the independent variables may or may not be continuous.

log phase of growth the stage of growth of microbial cells in which there is a constant exponential increase in numbers.

longitudinal study epidemiology study in which the exposed and nonexposed populations under investigation are identified at the beginning and followed concurrently throughout the calender year until the end of the study period; also termed concurrent prospective study or concurrent cohort study; in contrast to a retrospective cohort study.

lophotrichous having polar flagella arranged in a tuft.

lower limit of detection/lower detection limit (LDL) smallest quantity that can be accurately determined.

lower limit of quantitation/lower quantitation limit fewest number that can be accurately enumerated.

lower respiratory tract the trachea, bronchi, and lungs.

Lp1 *Legionella pneumophila* serogroup 1.

LPS lipopolysaccharide.

Lucibacterium bacterial genus; member of the Vibrionaceae; ability to luminesce.

luciferin-luciferase assay reduced luciferin reacts with oxygen to form oxidized luciferin in the presence of luciferase enzyme, magnesium ions, and adenosine triphosphate as an assay for estimating the amount of biomass present in a sample by assuming that the amount of light emitted is proportional to the concentration of adenosine triphosphate.

luminescence/luminescent production of light.

lunate crescent-shaped.

Lwoff and Tournier scientists who established a classification system for viruses based on the nature of the nucleic acid, the symmetry of the capsid, the presence or absence of an envelope, and the number of capsomeres or diameter of the nucleocapsid.

Lyme disease an acute inflammatory disease with a rash that may result in arthritis and heart and nervous system problems; caused by an infection with *Borrelia burgdorferi* resulting from a tick bite; first reported in Old Lyme, Connecticut.

lyophilization process in which cold temperature and air evacuation are used for preservation of microorganisms; also termed freeze-drying.

lyse/lysis rupture of a cell.

lysin a molecule that causes lysis.

lysogen a prokaryotic cell that contains a prophage.

lysogeny the ability of a bacteriophage to survive and be replicated through incorporation of its nucleic acid into that of its host, without causing cell lysis.

lytic causes lysis.

lytic infection the infection of a host cell by a virus in which, after progeny viruses are produced, the cell is lysed, thereby releasing newly synthesized virus particles.

lytic phage a virus that destroys its host bacterial cell; also known as virulent phage.

M

μ Greek letter mu used as the symbol for microbial growth (see Monod equation); formerly used as the symbol for micron which was replaced by micrometer (μm).

μm micrometer.

m^3 cubic meter; unit of measure describing the volume equal to 1000 liters.

MA muramic acid.

Mab monoclonal antibody.

MAC MacConkey's agar; *Mycobacterium avium* complex.

MacConkey's agar (MAC) a differential culture medium that incorporates neutral red to distinguish between lactose and non-lactose fermenting bacteria.

macroconidia plural of macroconidium.

macroconidium a single, large, multicellular conidium.

macrocyclic an organic molecule that contains a cyclic element with more than 15 atoms.

Macromonas bacterial genus; cylindrical to bean-shaped cells that oxidize sulfur and sulfur compounds; may accumulate calcium carbonate with sulfur globules; found in seawater.

macrophage long-lived phagocytic cell in mammals that can engulf bacteria and parasites, and may become activated in response to foreign materials resulting in the release of substances that stimulate other cells of the immune system.

macule a spot.

Madin-Darby bovine kidney cells (MDBK) cells from the kidneys of cattle that have been adapted to grow *in vitro* as a continuous cell line; used for the detection and cultivation of microorganisms such as *Cryptosporidium*.

magnetic stir bar stirbar.

magnetosomes membrane-bound, nanometer-sized particles of iron oxide found in cells that exhibit magnetotaxis.

Magnetospirillum bacterial genus; magnetotactic organisms that participate in biomineralization of magnetosomes.

magnetotaxis motility of microorganisms directed by a geomagnetic field.

magnification observed increase in size of an object when viewed through a microscope or lens.

MAIS group a grouping of environmental, opportunistic pathogens of animals and humans of mycobacteria frequently found in natural and drinking waters; *Mycobacterium avium*, *Mycobacterium intracellulare*, and *Mycobacterium scrofulaceum*.

male-specific a bacteriophage that attaches to the F-pilus of the host cell; F+RNA phage and F+DNA phage.

manganese cycle the transformation of manganese in reservoirs by biological activity; the oxidation of the manganous (Mn^{2+}) form to manganese dioxide (MnO_2) which is insoluble in water and the further oxidation to the manganic (Mn^{4+}) form by fungi and soil and aquatic bacteria, and the anoxic reduction of Mn^{4+} to Mn^{2+} by chemoorganotrophic organisms.

manifold a chamber with multiple channels or openings for the flow of liquid or air.

mannan a cell wall constituent in yeast; a mannose-containing polysaccharide storage material in plants.

Man-Rogosa-Sharpe medium de Man-Rogosa-Sharpe medium.

mar multiple antibiotic resistance locus present in *Citrobacter*, *Enterobacter*, *Escherichia coli*, *Hafnia*, *Klebsiella*, *Salmonella*, and *Shigella* that is also associated with tolerance to organic solvents, disinfectants, and weak acids.

marine snow macroscopic organic aggregates containing algae, zooplankton molts, organic and inorganic debris, and bacteria that are involved in recycling of nutrients in the open ocean.

masking interference that results in the inability to detect the analyte.

MATCI miniature thermal cycling integrated reaction chamber

material safety data sheet (MSDS) detailed information about a substance including the source, chemical properties, hazards identification, first aid and emergency response measures, handling, storage, and personnel protective equipment recommendations, health effects that may result from exposure, and disposal considerations; generally provided by the manufacturer to the user.

matrices plural of matrix.

matrix material surrounding or integrated with the item of interest.

maximum growth temperature the temperature above which a particular microorganism cannot grow; at this temperature protein denaturation occurs with a collapse of the cytoplasmic membrane and thermal lysis.

maxiprep colloquial term used to describe the purification of plasmids from bacterial suspensions of 100–500 ml in volume; in contrast to megaprep, midiprep, and miniprep.

McFeters chamber a dialysis unit constructed with plexiglass walls to provide rigidity used in the study of microorganisms in freshwater environments; the dialysis membranes allow water and dissolved substances to move freely through the unit.

MCYSTs microcystins.

MD mean deviation.

MD8 an impactor sampler that utilizes filtration for the collection of bioaerosols into a gelatin membrane.

MDBK cells Madin-Darby bovine kidney (MDBK) cells.

m-E agar selective culture medium used for the isolation of enterococci from water.

mean (μ) arithmetic mean.

mean deviation (MD) the average of the absolute deviation of data.

mechanical vector an animate object that serves to transfer a pathogen to a new host; the concentration of infectious agent does not increase while in contact with the vector.

media plural of medium.

median the statistical point where half of the individual data points are greater and half of the data points are lower in value; in contrast to arithmetic mean.

medium a semisolid or liquid growth matrix used in the study of microorganisms.

megaprep colloquial term used to describe the purification of plasmids from bacterial suspensions greater than 500 ml volumes; in contrast to maxiprep, midiprep, and miniprep.

melting the separation of double-stranded DNA into two single strands using heat; in contrast to denaturation.

MEM minimal essential medium.

membrane a thin layer that is permeable or semipermeable.

membrane filter a sieve matrix of cellulose acetate or cellulose nitrate manufactured as a surface of approximately 80–85% open holes for rapid filtration of liquid with the retained material enmeshed in the matrix; most commonly used method of filtration in microbiology; in contrast to a nucleopore filter.

membrane filter adsorption method used for the concentration of viruses in water samples that relies on the retention of particles on a filtration matrix and subsequent elution for analysis.

membrane filter (MF) technique standardized procedure specified in the Safe Drinking Water Act for the analysis of total coliform bacteria in drinking water by water utilities; the method involves the filtration of a 100 ml water sample onto a filter and incubation on eosin methylene blue (EMB) agar; results may not exceed 1 colony-forming unit/100 ml as the arithmetic mean of all samples examined per month, 4 CFU/100 ml in more than 1 sample when <20 samples are examined/month, or 4 CFU/100 ml in >5% of the samples when >20 samples are examined/month; results are reported to the United States Environmental Protection Agency and if the data do not meet the prescribed standard the water utility must notify the public and correct the problem.

M-Endo agar selective culture medium used for the isolation of total coliform bacteria from water during confirmatory water quality testing with a characteristic metallic green sheen viewed as a positive result, reddish colonies characteristic of total coliform bacteria, and colorless colonies characteristic of non-lactose fermenters.

meningitis an inflammation of the membranes surrounding the brain and spinal cord.

meningoencephalitis an inflammation of the brain and the membranes surrounding the brain and spinal cord.

mercury methylation the microbial conversion of mercuric ion (Hg^{2+}) to methylmercury (CH_3Hg^+); reaction occurs with Hg^{2+} ions that are absorbed on particulate matter, especially in aquatic environments, thereby forming soluble CH_3Hg^+, an environmental toxin that can be concentrated in the food chain or further metabolized to dimethylmercury (CH_3-Hg-CH_3) which is a volatile environmental toxin.

merosporangium chain of fungal spores that is formed by the simultaneous division of an elongated cell.

mesokurtic reference to the kurtosis of a graph when the data are normally distributed and therefore in the shape of a bell curve; in contrast to leptokurtic or platykurtic.

mesophile/mesophilic descriptive of the temperature requirements of microorganisms with an optimum growth temperature range of 25–40°C.

mesoscale spread dispersal of a crop pathogen from the initial site of infection extending over a large area of one field or over many fields up to several hundred kilometers in size or a continent during a single growing season; in contrast to microscale and synoptic (macroscale) spread.

mesotrophic moderately productive lake or surface freshwater resulting from an intermediate amount of nutrient.

meta-analysis statistical analysis of a large collection of results obtained from several individual studies with weighting of the data due to differences in the number of samples analyzed.

metabolism the collection of biochemical reactions within a cell that generate or require energy; generally classified as either anabolic (synthetic) or catabolic (breakdown) reactions.

metabolite the end product of metabolism.

methane oxidization process conducted by methanotrophs in soil, water, and landfills in which methane, methanol, halomethanes, and methylamines are oxidized and the carbon incorporated into cell biomass.

methanogen member of the Archaea that reduces CO_2 using H_2; strict anaerobic microorganism that has redox potentials ranging from −350 mV to −450 mV and is chemolithotrophic, using CO_2 as the sole carbon source with the first reaction resulting in the formation of methane.

methanogenesis conversion of CO_2 to methane, in contrast to acetogenesis.

Methanopyrus bacterial genus; gram-positive bacillus isolated from submarine hydrothermal vents; methanogen that produces methane from H_2 and CO_2; temperature optimum of 100°C.

methanotroph a microorganism that is capable of oxidizing methane with molecular oxygen and uses the CH_4 as a carbon and energy source; two families are recognized, Methylococcaceae and Methylocystaceae.

methoxychlor 1,1,1-trichloro-2, 2-bis(*p*-methoxyphenyl)ethane, a pesticide with *p*-methoxy groups that are subject to microbial dealkylation; in contrast to DDT.

methyl iodide (CH_3I) compound that is an effective carrier of iodine from the biosphere into the atmosphere produced by a variety of phytoplankton and marine bacteria.

Methylobacter bacterial genus; methanotroph; member of the family Methylococcaceae.

Methylococcaceae bacterial family; characterized as methylotrophs; member of the subclass of *Protobacteria*; members preferentially grow at low methane concentrations; associated with the rhizoplane of rice plants.

Methylocystaceae bacterial family; characterized as methylotrophs; member of the subclass of *Protobacteria*; members preferentially grow at high methane concentrations.

Methylomonadaceae bacterial family; characterized by their ability to utilize carbon monoxide, methane, or methanol as the sole carbon source.

Methylosinus bacterial genus; methanotroph; member of the family Methylocystaceae that outcompetes *Methylomonas* under nitrogen limiting conditions.

methylotroph microorganism that utilizes single carbon compounds such as methane, methanol formate, and carbon monoxide as their carbon source.

metric ton per annum (MTA) unit of measurement used in describing global annual production rates in biogeochemical cycles.

metula a small branch of a conidiophore that produces a phialide or a conidiogenous cell.

metulae plural of metula.

MF technique membrane filter technique.

mg milligram.

MIC microbial induced corrosion; minimum inhibitory concentration.

microaerophile/microaerophilic a microorganism that requires oxygen for growth but can tolerate concentrations that are lower than atmospheric conditions.

microarray placement of multiple single-stranded DNA segments on a miniaturized support surface such as a slide or computer chip.

microbial ecology the science that examines the relationships between microorganisms and the biotic and abiotic environment.

microbial fallibility principle formalized by Alexander in 1965 that no natural organic compound is totally resistant to biodegradation provided that environmental conditions are favorable.

microbial induced corrosion (MIC) the accelerated corrosion of a metal or an alloy by the activity of a consortium of microorganisms.

microbial pest control agent (MPCA) a nonpathogenic microorganism applied to agricultural crops to minimize the colonization of a microbial phytopathogen.

microbial volatile organic compounds (MVOCs) volatile organic compounds produced as a by-product of microbial metabolism; mushroom-derived compounds produce pleasant odors and flavors, but other fungal-derived compounds are associated with a musty, moldy odor; many fungal compounds are derivatives of alcohols, ketones, hydrocarbons, and aromatics; indoor concentrations are generally in the ng/m^3 range, but threshold limit values for these compounds have not been established; detection in indoor environments may indicate the presence of microbial contaminants.

microchip PCR technique utilizing computer chips and polymerase chain reaction amplification for the detection of microbial contaminants allowing for rapid screening of samples using thousands of oligoprobes.

microconidia plural of microconidium.

microconidium a small, generally unicellular, asexual cell; may function in sexual reproduction in some genera.

microcosm a small-scale experimental model that is designed to reproduce the environmental conditions of interest as closely as possible; used in laboratory experiments to define environmental conditions and test biological populations.

microcystins (MCYSTs) cyclic heptapeptide toxins that are produced by some cyanobacteria; currently more than 60 toxins are recognized, most cause potent inhibition of protein phosphatase 1 and 2A in plants and animals.

Microcystis aeruginosa bacterial species; a microcystin-producing freshwater cyanobacterium that blooms in eutrophic lakes and reservoirs in warmer seasons; associated with animal and human poisonings.

microecosystem microcosm.

microelectrophoresis method used to measure multiple electrophoretic mobilities of microorganisms in a liquid suspension placed into an electrically charged flow chamber; the velocity at which negatively charged organisms migrate to the positive electrode and the positively charged organisms migrate to the negative electrode are recorded.

microenvironment the physical and chemical conditions in the area immediately surrounding an organism.

micrometer (µm) unit of measure describing the length of 10^{-6} meter.

micron micrometer.

microorganism a microscopic organism that exists as a single cell or in an aggregate of cells, or as an acellular entity (i.e., virus).

microscale spread dispersal of a crop pathogen that is limited to a few hundred meters within one field and occurring during a single growing season; in contrast to mesoscale and synoptic (macroscale) spread.

microscope instrument used to magnify the size of the apparent image of an object; used for the observation of microorganisms and other cells, and details of their structure; the variety of microscopes and techniques include atomic force microscopy, light microscopy, scanning electron microscopy, transmission electron microscopy.

microscope slide glass surface used for the mounting of specimens for analysis by microscopy.

microscopic particulate analysis (MPA) analytic method developed for water utilities to determine whether a ground water supply should be classified as groundwater under the direct influence of surface water, and therefore regulated as such by the United States Environmental Protection Agency; method involves filtering a large volume (e.g., hundreds of liters) of water and examining the filtrate for indicators of surface water contamination (e.g., pollen, insect parts, diatoms, algae, and protozoan cysts).

microscopy use of a microscope for analysis.

microstats steady-state two-dimensional diffusion gradient devices used to study biofilms.

microtiter plate a plastic rectangular dish, generally containing at least 96 wells, that is used for experiments conducted with small volumes.

microtubule a structural entity of eukaryotic flagella.

midiprep colloquial term used to describe the purification of plasmids from bacterial suspensions of 10–100 ml volumes; in contrast to maxiprep, megaprep, and miniprep.

mid-spread interquartile range.

mildew fungi associated with agriculture (e.g., powdery mildew and downy mildew); informal terminology used to describe the dark staining biological film present in moist indoor areas, especially in the bathroom (e.g., the grout of shower tiles and plastic shower curtains).

milligram (mg) unit of measure describing the weight of 10^{-3} grams.

Millipore filter commercial filtration product.

Milli-Q water ultrapure water produced by a series of processes, including filtration, for a number of technical applications, including preparation of culture media, chromatography, and pharmaceutical applications.

minimal essential medium (MEM) a growth matrix prepared with the only basic nutrients needed to support growth; a solution containing essential nutrients, including amino acids and vitamins, that is used as a growth and maintenance medium in cell culture systems; it is often supplemented with antibiotics to prevent the growth of bacteria and fungi, fetal calf serum, and buffers.

miniature thermal cycling chamber integrated (MATCI) reaction chamber a micro-machined silicon high-efficiency device for real-time fluorescence monitoring of product DNA.

minimum growth temperature the temperature below which a particular microorganism cannot grow generally due to the occurrence of gelling in the cytoplasm and diminished transport within the cell.

minimal medium minimal essential medium.

minimum inhibitory concentration (MIC) the lowest concentration required to prevent microbial growth.

miniprep colloquial term used to describe the purification of plasmids from bacterial suspensions of 1–10 ml volumes; in contrast to maxiprep, megaprep, and midiprep.

miscible liquids that are soluble in all proportions.

misclassification bias the introduction of a systematic error due to inaccurate measurement tools; in contrast to surveillance bias and selection bias.

mist a suspension of airborne liquid droplets that are generally 2–50 μm in diameter.

mixotroph/mixotrophic an organism that requires an inorganic compound as an electron donor and an organic compound as a carbon source.

mmHg millimeters of mercury, used to express atmospheric pressure.

m^3/min rate measurement of the volume of air or liquid per unit time; 1 cubic foot/minute = 28 liters/min = 0.028 m^3/min.

mode statistical interpretation for nominal data as the value that is most frequently observed; if two categories are of similar value the term bimodal and if more than two categories are of similar value then the term multimodal is used.

modeling use of mathematical equations to represent a situation; often used to predict the outcome of an event.

MOI multiplicity of infection.

moisture meter instrumentation used to detect moisture in building materials; two general types are currently used, a noninvasive device that uses electrical impedance, and a probe/pin type meter that utilizes electrical conductance between two probes which are inserted into the material being tested.

mold common term for filamentous fungi.

molecular beacon nucleic acid probe with a stem and loop structure containing a fluorescent moiety on one arm of the stem and a quencher on the other arm; when a complementary target sequence is encountered, the probe spontaneously undergoes a conformational change, forcing the arms of the stem to separate, and allowing fluorescence to occur.

molecular weight cutoff (MWCO) the lower limit of molecular weight that is retained, for example, on a filter.

molecular weight marker a mixture of different nucleic acid fragments of known molecular weight that is used to calibrate the nucleic acid fragments in a sample after separation by gel electrophoresis.

molecule more than two atoms bonded together.

Mollicutes microbial class; fundamental characteristic of the microorganisms in this grouping is the absence of a cell wall due to an inability to synthesize peptidoglycan; generally gram-negative; subdivided into the order Mycoplasmatales and several families including Mycoplasmataceae and Acholeplasmataceae.

moniliaceous in a chain or a series of segments.

monitoring the collection of data to characterize specific conditions during a defined time period.

monoclonal antibody (Mab) an antibody that is produced from a single clone of cells; in contrast to polyclonal antibody.

Monod equation of bacterial growth an equation that expresses the relationship between the culture generation time and the concentration of the limiting substance.

$$\mu = \frac{\mu_{max} S}{K_s + S}$$

monolayer a single layer; a single layer of cells adhering to a solid surface such as a culture flask.

monopodial having a single pseudopod; used to describe amoeba; in contrast to polypodial.

morbidity illness.

mordant chemical used to fix a color or dye.

morphology shape.

mortality death.

mortality rate the number of individuals that die of a disease divided by the number of individuals in the population; in contrast to case-fatality rate.

most-probable number (MPN) standardized procedure for the estimation of microbial concentrations based on the assumption that microorganisms are dispersed in the liquid sample according to the Poisson distribution.

most-probable number cytopathogenic unit (MPNCU) the number of infectious viruses present in a sample as estimated using the most probable number procedure.

motile capable of moving oneself.

mottling colored spots or blotches.

MPA microscopic particulate analysis.

MPCA microbial pest control agent.

MPN most-probable number.

MPNCU most-probable number cytopathogenic unit.

mRNA messenger RNA.

MRS de Man-Rogosa-Sharpe medium culture medium for growth of lactobacilli.

MS2 an icosahedral F+RNA phage that is the type strain for serological group I of the RNA phages; commonly used as an indicator for enteric viruses due to its similar size, shape, and resistance to environmental stresses.

MSDS material safety data sheet.

MTA metric ton per annum.

m-Tec agar selective culture medium for the isolation of *Escherichia coli* from water samples.

mu Greek symbol used to denote arithmetic mean.

Mu a temperate virus that is also a transposable phage and a mutator phage used in genetic engineering.

mucedinoid musty or moldy odor.

mucilage sticky mixture of carbohydrates found in plants.

mucilaginous slimy.

Mucor fungal genus; woolly colony on malt extract agar that is grayish to brown on the surface and pale on reverse; rhizoids and stolons are not present; round to ellipsoidal sporangiospores that are borne on non-apophysate sporangia and columella; isolated from organic matter, dung, soil, leftover food, soft fruit, and juices; water activity of 0.09–0.94; associated with Type I allergies and Type III hypersensitivity; ubiquitous, several species currently recognized; used in production of cheese.

Mucorales an order within the class Zygomycetes of rapid growing fungi with large aseptate or rarely septate hyphae and sporangia, examples of which are *Mucor* and *Rhizopus*.

mucous membrane secreting membranes lining the hollow organs of the body, such as the eyes, nose, and mouth.

mucous membrane irritation syndrome an assortment of symptoms including nasal stuffiness, sinus headaches, and decrease of smell in response to inhalation exposure to bioaerosols.

multimodal data in which more than two peaks are observed.

multipipette hand-held device capable of housing several disposable tips for transfer of a specified amount of liquid; generally used for transfer of liquids in a microtiter plate.

multiple correlation coefficient (R) statistical value used with multiple regression analysis; this value squared is the coefficient of determination.

multiple regression statistical analysis used to illustrate the relationship between one dependent variable and more than one independent variable all of which are interval data; in contrast to simple regression.

multiplex PCR a variation of the polymerase chain reaction amplification method in which two or more sets of primers are used to amplify different targets in a single reaction.

multiplicity of infection (MOI) the average number of virus particles per cell.

municipal drinking water standard the maximum allowable concentration of a contaminant allowed to be present in public drinking water; established by the United States Environmental Protection Agency.

Muntz, A. reported microbial nitrifying activity in 1877–1879 with Schloesing; demonstrated that the ammonium in sewage was oxidized to nitrate when passed through a sand column, an activity that was eliminated when chloroform was introduced and then restored when a soil suspension was used as an inoculum.

muramic acid (MA) chemical compound present in the cell wall of most bacteria; used to estimate microbial biomass with an accepted ratio of 44 μg muramic acid per mg of carbon for gram-positive bacteria and an accepted ratio of 12 μg muramic acid per mg of carbon for gram-negative bacteria.

murein peptidoglycan.

muriform structures that are transversely and longitudinally septate.

mushroom the fruiting structure of a basidiomycete; some produce polypeptide or amino acid-derived compounds that are toxic with an ingestion route of exposure.

mutacins antimicrobial substances produced by some strains of *Streptococcus mutans*.

mutagen/mutagenic capable of altering the genetic material in a living cell.

mutation an inheritable change in the genes of a genome; a change in the nucleotide order of a nucleic acid.

mutualism a highly specific and physically close obligatory relationship between microbial populations that benefits both populations; also called symbiosis.

MVOC microbial volatile organic compound.

MWCO molecular weight cutoff.

mycelia plural of mycelium.

mycelia sterilia sterile mycelia.

mycelium the mass of hyphae that is the vegetative form of a fungus.

Mycobacteriaceae bacterial family; taxonomically in the order Actinomycetales; characterized as acid fast cells that do not form mycelia or spores; representative genus is *Mycobacterium*.

Mycobacterium bacterial genus; gram-positive, pleomorphic cells that may exhibit branching or filamentous growth; the presence of mycolic lipids in the cell wall result in the acid fast staining of members of this genus.

Mycobacterium avium bacterial species; flat, translucent, nonpigmented colonies that may appear yellow with age; causative agent of tuberculosis in fowl.

***Mycobacterium avium* complex (MAC)** term used to describe *Mycobacterium avium* and *Mycobacterium intracellulare*; causes disseminated infections in immunocompetent individuals and persons with AIDS; infection may result in pulmonary disease; route of transmission is unknown, although environmental routes, including water, are suspected.

Mycobacterium kansasii bacterial species; photochromogenic colonies when cultured in the light; somewhat rough colonies that are readily emulsified in water; pathogen to humans causing pulmonary lesions but not contagious person-to-person; dermal route of exposure through breaks in the skin may result in subcutaneous lesions.

Mycobacterium marinum bacterial species; smooth to rough, photochromogenic colonies; infection generally in the form of skin lesions resulting from the dermal route of exposure through abrasions and contact in swimming pools, fish and aquariums.

Mycobacterium tuberculosis bacterial species; slow-growing, nonpigmented, human pathogen that is transmitted via the inhalation route of exposure that may result in tuberculosis, a respiratory disease, depending on the virulence of the organism and host resistance.

mycobiont the consumer population in a mutualistic relationship.

Mycocystis aeruginosa a toxigenic blue-green algae that proliferates in fresh water.

mycologist a scientist specializing in the study of fungi.

mycology the study of fungi.

Mycoplasma microbial genus; member of the family Mycoplasmataceae with approximately 50 recognized species; absence of a cell wall associated with the inability to synthesize peptidoglycan results in instability in morphology, susceptibility to lysis by detergents and alcohol, osmotic sensitivity, and resistance to penicillin; considered the smallest organisms that are capable of self-reproduction; some species cause disease in plants and animals.

Mycoplasmataceae microbial family; characteristics include a requirement of sterol for growth and NADH oxidase localized in the cytoplasm.

mycorrhiza/mycorrhizae mutualism in which a fungus becomes integrated into the physical structure of a plant root with the fungus deriving nutritional benefit and not causing disease.

mycorrhizal fungi the fungi involved in mycorrhizae.

mycoses plural of mycosis.

mycosis infectious disease caused by pathogenic fungi; categorized as endemic or opportunistic.

mycotoxin naturally occurring secondary metabolite produced by some fungi during growth; currently hundreds are known; chemically belong to the alkaloids, cyclopeptides, and coumarins; generally low molecular weight, low volatility; possible ingestion, inhalation, or dermal route of exposure, depending on the specific compound and exposure situation.

myeloma a malignant tumor of antibody-producing cells.

myocarditis an inflammation of the heart muscle.

Myxobacterales a group of gliding bacteria characterized as small bacilli that are normally embedded in a slime layer that can form fruiting bodies; generally found on decaying plant material, animal dung, or the bark of living trees, and their hydrolytic enzymes may lyse other microorganisms.

Myxomycetes fungal class; slime mold; within the Kingdom Protista; considered to have characteristics similar to the true fungi and animals; airborne spores detected using microscopic assay of spore trap air samples difficult to distinguish from the teliospores of smuts; may elicit allergic reactions.

N

n, N the number of subjects.

N6 Andersen single-stage impactor sampler.

na not applicable.

NA numerical aperture.

naked virus a virus without an envelope surrounding the capsid.

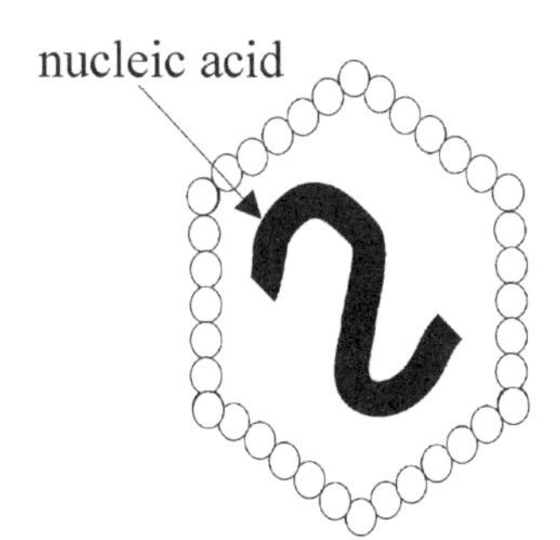

nalidixic acid antimicrobial agent that causes a reversible inhibition of bacterial DNA synthesis and is used in the Kogure technique to determine the viability of cells in suspension.

nanobacteria smallest cell-walled bacteria currently recognized.

nanotechnology development of miniaturized devices to reduce the size, weight, and reagent volume requirements of biotechnology instrumentation.

NAS National Academy of Sciences.

National Academy of Sciences (NAS) a private, nonprofit society of distinguished scientists and engineers that advises the federal government on scientific and technical issues; granted a charter by congress in 1863.

National Collection of Industrial and Marine Bacteria Ltd. (NCIMB) culture collection entity located in Aberdeen, Scotland that catalogs and sells standard strains of microorganisms.

National Institute of Occupational Safety and Health (NIOSH) an institute of the Centers for Disease Control and Prevention; the agency of the United States government that is responsible for conducting research and providing recommendations for the prevention of occupational injury and disease, and provides training to occupational safety and health professionals.

National Institutes of Health (NIH) founded in 1887, the NIH is one of eight health agencies of the U.S. Public Health Service, and is responsible for conducting and directing medical research in the United States.

National Research Council (NRC) organized by the National Academy of Sciences in 1916 to provide services to the government, public, and the engineering and scientific community; administered by the National Academy of Sciences, the National Academy of Engineers, and the Institute of Medicine through the National Research Council Governing Board.

natural turnover time a measurement of secondary productivity as the amount of time needed for an existing heterotrophic microbial population to take up and/or respire a quantity of substrate equal in concentration to the existing *in situ* concentration.

Naumanniella bacterial genus; bacilli that oxidize iron and are encapsulated with iron oxides; widely distributed in iron-rich waters.

NB nutrient broth.

NCIMB National Collection of Industrial and Marine Bacteria Ltd.

necrosis death of cells or tissue; also called rot in plant pathology.

negative feedback mechanism in which the products of a multistep process react on an earlier step to prevent the formation of that product; in contrast to positive feedback.

negative predictive value the ability of a test method to correctly identify true negatives; calculated by dividing the number of true negatives by the total number of negatives, a value that includes both the true negatives and the false negatives.

negative skew the shifting of the symmetry of a curve to the left; see skew.

negative stain the use of a dark dye such as aqueous nigrosin to stain background material and permit the unstained microbial elements to be viewed with light microscopy; also used to describe the use of a fluorescein solution that results in the image of organisms as dark objects against a bright background.

Neisseriaceae bacterial family; characterized as gram-negative cocci and coccobacilli; two members of this family are significant human pathogens, *Neisseria gonorrhoeae* (causative agent of gonorrhea) and *N. meningitidis* (a causative agent of bacterial meningitis).

NBOD nitrogenous biochemical oxygen demand.

neoplasm new or abnormal growth, such as a tumor.

Neotyphodium fungal genus; an endophytic symbiont of grasses that confers characteristics to the plant such as resistance to herbivores, drought resistance, and enhanced growth in exchange for plant-provided nutrients.

nephelometric turbidity unit (NTU) unit of measurement for turbidity.

nephrotoxic harmful to the kidney.

neritic zone region of ocean water near the shore where the depth is less than 200 meters.

nested PCR a polymerase chain reaction amplification technique in which two primer sets are used sequentially, an outer set producing a product that is then subjected to amplification with the inner set.

nested primers primer sets selected such that one set amplifies a region located within that amplified by the preceding set.

nested variables statistical condition in which each variable occurs at only one level.

net community productivity the net gain in organic matter produced by photosynthesis in the carbon cycle and not converted back to CO_2.

neural relating to the nervous system.

neurotoxic harmful to the nervous system.

neurotrophic having an affinity for nerve tissue.

neuston the surface microlayer that is the uppermost layer of the hydrosphere as an interface between the hydrosphere and the atmosphere; a high surface tension region that is favorable for photoautotrophic organisms due to light, CO_2 from the atmosphere, and enrichment of nutrients and metals, and favorable for secondary producers due to the presence of O_2 from the atmosphere and accumulation of organic compounds; freshwater equivalent of the pleuston.

neutralism a relationship of microbial populations in which there is no interaction between the populations; occurs when populations of organisms have different metabolic capabilities, populations are spatially distant from each other, when environmental conditions are unfavorable for active growth, or when organisms are in a resting state.

neutrophile a microorganism that prefers neutral pH (~7.0).

nexus a link or connection.

NHE nonhemolytic enterotoxin.

niche the functional role of an organism within an ecosystem; a description of not only the location of an organism but what the organism does at that location.

Niel, C.B. van van Niel, C.B.

nigrosin a solution used as a negative stain when prepared as an aqueous solution.

Nigrospora fungal genus; wooly colony on malt extract agar that is white becoming black on the surface and reverse; the smooth walled, ovoid to ellipsoidal, unicellular conidia are black with a slight oblate on the horizontal and have an equatorial germ slit; active spore discharge.

NIH National Institutes of Health.

NIOSH National Institute of Occupational Safety and Health.

nitrification the conversion of ammonia to nitrite and then nitrate resulting from microbial activity.

nitrifying bacteria a group of bacteria that oxidizes ammonia to nitrite or nitrite to nitrate.

Nitrobacter bacterial genus; member of the Nitrobacteraceae; gram-negative bacilli that oxidize nitrite to nitrate; isolated from soil, freshwater and marine environments, and are unusual in their ability to grow slowly on acetate or pyruvate as a sole carbon and energy source.

Nitrobacteriaceae bacterial family; characterized as nitrifying bacteria; commonly isolated from soil, freshwater, and marine environments.

nitrocellulose material that has a high affinity for macromolecules and is used in filter paper and beads for purification of biological samples.

Nitrococcus bacterial genus; member of the Nitrobacteraceae; gram-negative cocci that oxidize nitrite to nitrate; found in the Pacific Ocean.

nitrogen cycle a series of chemical and biochemical reactions that effect the transformation of nitrogen in reservoirs.

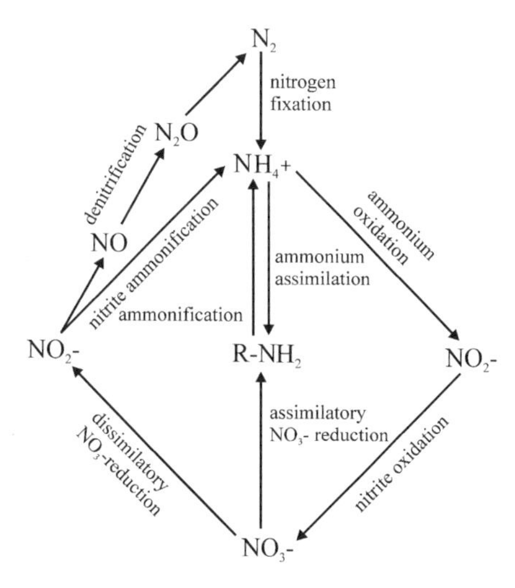

nitrogen fixation the microbial-mediated conversion of molecular nitrogen, N_2, to organic nitrogen.

nitrogenous biochemical oxygen demand (NBOD) amount of oxygen consumed by microorganisms for the biochemical oxidation of inorganic nitrogenous compounds.

nitrogen stripping removal of nitrogen by volatilization of ammonia, NH_3, at high pH.

***o*-nitrophenyl-β-galactopyranoside (ONPG)** a chromogenic substrate used to detect β-D-galactosidase produced by coliform bacteria.

Nitrosococcus bacterial genus; member of the Nitrobacteraceae; gram-negative cocci that oxidize ammonia to nitrite; widely distributed in seawater.

Nitrosolobus bacterial genus; member of the Nitrobacteraceae; gram-negative pleomorphic and lobate cells that oxidize ammonia to nitrite; isolated from some soils.

Nitrosomonas bacterial genus; member of the Nitrobacteraceae; gram-negative ellipsoidal or short bacilli in pairs or short chains that oxidize ammonium to nitrite; widely distributed in soils.

Nitrosospira bacterial genus; member of the Nitrobacteraceae; gram-negative spiral-shaped cells that oxidize ammonia to nitrite; widely distributed in soils.

Nitrospina bacterial genus; member of the Nitrobacteraceae; gram-negative slender, straight bacilli that oxidize nitrite to nitrate; found in the Atlantic Ocean.

NMR nuclear magnetic resonance.

NOAEL no observable adverse effect level.

noble rot *Botrytis cinerea*.

Nocardiaceae bacterial family; taxonomically in the order Actinomycetales; characterized as saprophytic or facultative parasites that produce mycelia that fragment into nonmotile cells, and spores that are not borne in sporangia but aerial spores are usually absent; representative genera include *Nocardia* and *Pseudonocardia*.

node a joint, point of origin of fungal hyphae, or an enlarged area on a fungal hypha.

nod factors oligosaccharides that induce root hair curling and trigger the division of plant cells resulting in the formation of root nodules by nitrogen-fixing bacteria.

nod genes genetic sequences in nitrogen-fixing bacteria that direct specific steps in the formation of a root nodule.

nodular body a round, knot-like formation of intertwined hyphae.

Nodularia bacterial genus; filamentous, heterocystous nitrogen-fixing cyanobacteria isolated from soils, plants, and brackish and fresh waters; organisms form toxic blooms in brackish waters with the production of nodularin.

nodularin a cyclic polypeptide hepatotoxin.

nodulate having intermittent areas of thickening or nodes.

nodule a tumor-like structure in legume plant roots that contains the microorganisms that participate in nitrogen-fixation activity.

NOEL no observable effect level; no observable adverse effect level.

nominal data/ nominal variables data that are expressed as a category without an implied order; a property either exists or it does not exist; in contrast to ordinal data/ordinal variables.

noncommunity water system a drinking water supply that serves an average of at least 25 individuals for 60 or more days per year; see also transient noncommunity water system and nontransient noncommunity water system.

nonculturable state in which a microorganism is viable but does not form a colony on agar-based growth medium under laboratory conditions.

nonenveloped refers to a virus particle that is not surrounded by a membranous envelope.

nonlantibiotic a class II, unmodified bacteriocin.

nonparametric estimators a statistical approach in which the proportion of species that have been identified before are compared to those that have been observed only once.

nonpolar hydrophobic; not readily dissolved in water.

nonpermissive host cell a host cell in which transformation is accomplished but viral DNA cannot be replicated and it is incorporated into the host DNA; in contrast to permissive host cell.

nonsporulating mycelia sterile mycelia.

nonstarter lactic acid bacteria (NSLAB) a variety of lactic acid bacteria that proliferate during ripening of cheese that may have a role in flavoring but also can result in defects in the products, such as the formation of calcium lactate crystals and cracks.

nontransient noncommunity water system a noncommunity water system that serves a continuous population of people a small portion of their water supply; examples include schools and businesses.

nonviable state in which a microorganism is not capable of reproduction.

no observable adverse effect level (NOAEL) a term used in dose-response studies to describe the highest dose of a substance given to the test population that does not produce an observable effect on the test population.

normal distribution Gaussian distribution.

normal flora microorganisms that are routinely present on or in another organism.

Northern blot hybridization procedure in which RNA is in the gel and the DNA or RNA is the probe; in contrast to the Southern blot hybridization method.

Norwalk-like agents term used to describe any one of a number of infectious agents that are serologically related to but distinct from the Norwalk virus; examples include the Hawaii, Snow Mountain Taunton, Moorcroft, Barnett, and Amulree agents; their serological relationships have not yet been determined.

Norwalk virus a small (27–32 nm) single-stranded RNA virus in the Caliciviridae family; transmitted by the fecal-oral route, it causes acute gastroenteritis (vomiting is a more common symptom than diarrhea) in humans following exposure to contaminated food or water; may be the etiologic agent responsible for the majority of cases of food- and waterborne gastroenteritis of whose etiology is currently unidentified.

nosocomial hospital-acquired.

NPV nucleopolyhedrovirus.

NRC National Research Council.

NSLAB nonstarter lactic acid bacteria.

NTU nephelometric turbidity unit.

nuclear magnetic resonance (NMR) analytical technique in which atoms are subjected to oscillating magnetic energy; used to identify the composition of organic compounds in a sample.

nucleic acid linear polymer of nucleotides linked by 3′,5′-phosphodiester linkages.

nucleocapsid a structure comprised of the nucleic acid and capsid of a virus.

nucleopolyhedrovirus (NPV) virus; a type of baculovirus.

nucleopore filter a polycarbonate film etched with chemicals to produce uniform holes of a defined pore size arranged vertically through the film; in contrast to a membrane filter.

nucleotide/nucleotide base the basic subunit of DNA or RNA; composed of a phosphate molecule, a purine or pyrimidine base [adenine, cytosine, guanine, thymine (in DNA), or uracil (in RNA)], and a sugar molecule (deoxyribose in DNA, ribose in RNA).

nucleus membrane-bound structure of eukaryotic organisms that contains DNA arranged in chromosomes.

null hypothesis (H_0) an expression that is used when developing an experimental design that states there is no difference between sets of data.

numerical aperture (NA) a measure of the light gathering ability of the objective lens of a microscope.

nutrient chemical used by an organism for metabolism.

nutrient agar a growth medium used for organisms with relatively simple nutrient requirements; composed of a source of amino acids and fatty acids (peptone) and a source of vitamins, carbohydrates, salts, and organic nitrogen compounds (beef extract); the addition of agar to the medium allows the material to be used in a semisolid form, such as in a petri plate.

nutrient broth (NB) a growth medium used for organisms with relatively simple nutrient requirements; composed of peptone (a source of amino acids and fatty acids) and beef extract (a source of vitamins, carbohydrates, salts, and organic nitrogen compounds).

nutristat an automated chemostat that provides a gradual increase in nutrient or contaminant used as a substrate for growth.

O

OA occupational asthma.

OAF open air factor.

obclavate an inverted club shape with the larger portion located at the base.

obconic conical-shaped with the broad end outward of the narrower base.

objective analysis determination by an analyst using a measurement methodology; in contrast to subjective analysis.

objective lens component of the compound light microscope that first magnifies the specimen, generally providing 10—100 times magnification.

obligate required condition.

obligate aerobe a microorganism that requires the presence of oxygen.

obligate anaerobe a microorganism that cannot grow in the presence of oxygen.

obligate intracellular parasite an infectious organism that is capable of reproduction only within a metabolizing host cell.

obovate/obovoid egg-shaped with the narrow portion at the base and the broad end pointed outward.

obpyriform pear-shaped with the broad end at the base.

occluded closed or blocked.

occupational asthma (OA) a reversible airway obstruction and/or increased bronchial responsiveness following exposure to a specific agent in the workplace.

Occupational Safety and Health Administration (OSHA) an office within the U.S. Department of Labor, created in 1970, whose mission is to ensure safe and healthful workplaces in America.

ochratoxin a mycotoxin produced by *Aspergillus ochraceus* and *Penicillium verrucosum* that causes neurological problems following an ingestion route of exposure.

Ochrobium bacterial genus; iron-oxidizing ellipsoidal to rod-shaped cells that are surrounded by a capsule containing iron oxides; distributed in iron-bearing fresh water.

ocular lens the eyepiece of the compound light microscope that provides an intermediate, inverted image from that of the specimen and adds an additional 10—15 times magnification.

OD optical density.

odds ratio the probability that an event will occur divided by the probability that an event will not occur; the ratio of the odds that an exposed population will develop disease divided by the odds that a nonexposed population will develop the disease such that if the exposure is related to the disease the resulting value will be greater than one, if the exposure is not related to the disease the value will equal one, and if the exposure is negatively related to the disease the value will be less than one.

odor character terms used to describe odors, such as fishy, rancid, acrid, or sweet.

odor threshold lowest concentration of a substance that can be detected by smell.

ODTS organic dust toxic syndrome.

OFAGE orthogonal field alternation gel electrophoresis.

O horizon region of the soil above the mineral soil that contains the organic matter formed from plant and animal materials; subdivided into the O_1 horizon characterized as the region where the plant and animal material is recognizable, and the O_2 horizon where the plants and animals have decayed.

oidium a conidium formed by the breaking of hypha.

Oidium fungal genus; common phytopathogen of leaves, stems, flowers and fruits; observed on spore trap air samples; commonly referred to as "powdery mildew."

oil-immersion lens a high numerical aperture objective lens that improves resolution in the compound light microscope; requires the placement of a drop of high grade optical oil between the specimen and the lens.

oleaginous oily.

olfactory related to the sense of smell.

oligo oligonucleotide.

oligonucleotide a short (generally, up to 20) sequence of nucleotides.

oligonucleotide probe an oligonucleotide used to identify a specific nucleic acid sequence; the probe hybridizes with a specific mRNA, and the presence of the hybridization product is detected using an assay such as dot blot or Southern hybridization.

oligoprobe oligonucleotide probe.

oligotrophic low carbon or low nutrient content; in contrast to copiotrophic.

olivaceous olive in color.

OM outer membrane.

omp outer membrane protein.

one-tailed test comparison of any difference between groups when the direction of the difference is specified in advance; used to test a directional hypothesis; in contrast to a two-tailed test.

one-way analysis of variance/one-way ANOVA statistical analysis used to compare multiple groups in a single test without considering interactions between the factors; in contrast to factorial ANOVA.

ONPG *o*-nitrophenyl-β-galactopyranoside.

oocyst a propagule of some parasites, notably the Coccidia and malarial parasites; generally resistant to environmental and chemical stresses.

Oomycetes aquatic fungi associated with fish diseases and potato blight; commonly referred to as water mold.

open air factor (OAF) the presence of pressure fluctuations, ions, relative humidity, and airborne pollutants, such as ozone and olefins that result in adverse effects to the survival of airborne microorganisms; targeted sites on the microorganism include phospholipids, proteins, and nucleic acids.

operational taxonomic unit (OTU) the specific identifier used to classify a subject as unique, such as the number of 16S rDNA similarity groups or the number of unique terminal restriction fragments.

operculate having a lid; in contrast to inoperculate.

operon a sequence of DNA that contains one or more structural genes plus regulatory genetic elements.

opportunistic pathogen a microorganism that usually resides in a commensal relationship with its host, but can cause disease in a debilitated host, often the result of immunosuppression, antibiotic regimen, or stress of the host.

optical density (OD) a measure of the density of a liquid determined by the absorption of light.

optimum growth temperature the temperature at which growth of a particular microorganism is the most rapid; at this temperature enzymatic reactions are occurring at the maximum possible rate.

orbicular circular outline.

order taxonomic level in the classification of bacteria below division includes several families.

ordinal data/ordinal variables data that consist of an ordered category where the differences between categories cannot be considered equal; often used to describe a patient's medical condition or a nonnumeric description of environmental quality; in contrast to an interval variable.

ordinate y-axis.

organelle a structure enclosed with a membrane, generally found in the cytoplasm of the cell.

organic compound a chemical that contains carbon; in contrast to inorganic compound.

organic dust toxic syndrome (ODTS) a systemic, nonallergic, noninfectious irritant response involving an acute inflammatory lung reaction resulting from inhalation exposure to dust containing endotoxin and fungal dust containing β-(1→3)-D-glucan or concentrations of greater than 10^9 spores/m^3 and may contain mycotoxin.

orthogonal field alternation gel electrophoresis (OFAGE) gel electrophoresis technique in which the electrical fields are at right angles to each other.

OSHA Occupational Safety and Health Administration.

osmolyte intercellular organic compounds that are compatible with cellular metabolism at high internal concentrations and are transported into the cell or synthesized within the cell to counterbalance external osmotic strength and serve as an osmoprotectant.

osmophile/osmophilic a microorganism that can tolerate high sugar or salt concentrations.

osmoprotectant a material that provides protection from stress of osmotic pressure.

osmosis diffusion of water across the cell membrane from a region of low solute concentration (high water concentration) to a region of high solute concentration (low water concentration).

osmotic pressure the force exerted from differences in solute concentration on opposite sides of a semipermeable membrane.

osmotic shock the bursting of a cell due to osmosis.

O-specific polysaccharide a compound that defines the antigenic uniqueness for region I of the lipopolysaccharide and thereby provides serologic specificity to the serogroups of gram-negative bacteria; responsible for quality of toxicity of different endotoxins.

ostiole an apical opening associated with fungal perithecia and pycnidia.

OTU operational taxonomic unit.

outbreak occurrence of a large number of cases of disease in a short period of time with a common source; for microorganisms, generally two individuals infected as a result of exposure to microorganisms from a common source must be identified; in the case of chemical poisoning or laboratory-confirmed primary amebic meningoencephalitis only one ill person is required for the incident to be classified as an outbreak.

outer membrane the layer that contacts the surrounding environment.

outgrowth the final stage of germination of endospores involving visible swelling, the uptake of water, and the synthesis of DNA, RNA, and protein.

outlier a value that is inconsistent with the other data obtained during statistical analysis.

oval egg-shaped with the broad end at the base.

ovate oval.

ovoid oval.

oxic containing oxygen.

oxidase reaction a biochemical test that is used to determine whether the organisms of interest contains cytochrome *c*, a protein involved in respiration.

oxidation a chemical or biochemical reaction in which the reactant loses electrons.

oxidation-reduction potential redox potential.

oxidizing agent a chemical that produces oxygen and gains electrons during a reaction.

oxyduric tolerant of oxygen.

oxygen cycle the transformation of oxygen in reservoirs by biological activity with oxygen being the preferred electron acceptor for aerobic and facultatively anaerobic organisms.

oxylabile killed or inactivated by the presence of oxygen.

ozonation the use of ozone to treat water, air, or food products to reduce the concentration of microbial populations.

ozone triplet oxygen (O_3), a strong oxidizing agent.

P

π Greek letter pi used as a symbol to denote geometric mean.

P-A presence-absence coliform test.

PAB propionic acid bacterium.

packaging the process in which viral genetic material is encapsulated by coat proteins.

PADs phenolic acid decarboxylases.

Paecilomyces fungal genus; colony on malt extract agar is white, pink/violet or olivaceous brown, but never blue or green; conidia produced in long chains from slender tapering and divergent phialides; ubiquitous; approximately 30 species; closely related to *Penicillium*; loose spores may be confused with those of *Aspergillus* and *Penicillium*; isolated from soil, decaying plant material, oak, composts, legumes, cigar tobacco, grapes, cottonseeds, jute fibers, leather, paper, polyvinyl chloride piping, fruit juice and bottled fruits, and optical lenses; water activity of 0.79–0.84; production of mycotoxin including paecilotoxins, patulin, and others.

Paecilomyces variotii fungal species; yellow-brown colonies; thermotolerant fungus that produces a sweet aromatic odor; Aw = 0.84; deteriorates jute fiber and paper and causes soft rot of timber.

PAGE polyacrylamide gel electrophoresis.

PAHs polyaromatic hydrocarbons.

paired t-test statistical analysis in which data are compared between matched subjects.

palindrome a sequence of DNA that is linked to the same sequence in reverse order.

pandemic a worldwide epidemic.

panduriform fiddle-shaped.

paniculate branched.

pantocins antibiotics produced by *Pantoea agglomerans* that are active against other bacteria.

Pantoea agglomerans bacterial species; possible use as a biological control agent to protect fruit trees from fire blight; synonym of *Erwinia herbicola*.

papillomavirus a genus of viruses in the family Papovaviridae; cause warts; some may be associated with the initiation of carcinoma.

Papovaviridae a family of nonenveloped, icosahedral viruses containing double-stranded DNA; human viruses in this family include the papillomaviruses and the polyomaviruses.

papovavirus any virus in the family Papovaviridae.

parabens esters of 4-hydroxybenzoic acid that are used as antimicrobial agents in food and cosmetic products and the pharmaceutical industry because of low toxicity, stability over a wide pH range, and activity against a broad spectrum of microorganisms.

Paracoccidioides brasiliensis fungal species; a diphasic fungus with a smooth to cerebriform yeast form of single (10–25 μm) to budding (1–10 μm) cells present at 37°C and a mycelial form at 25–30°C with a heaped, glabrous or wrinkled colony cultured on laboratory media showing short, white aerial mycelia that turn brown with age; some oval to round conidia may be observed on culture media, but these are indistinguishable from *Blastomyces dermatitidis* requiring conversion to the yeast stage for identification; causative agent of paracoccidioidomycoses.

paracoccidioidomycoses chronic infection of the respiratory system resulting from inhalation of *Paracoccidioides brasiliensis*.

Parafilm opaque wrap with an elastic property that is often used in the laboratory to seal containers to minimize leaking or evaporation.

parametric estimators a statistical approach in which the number of unobserved species in a community are estimated by fitting sample data to models of relative species abundance.

parasite an organism that lives on or in a host.

parasitism relationship of microbial populations in which one population derives its nutritional requirements at the harm of the other population.

partial correlation statistical estimate of the correlation of each variable with the dependent variable after all other variables have been considered.

partial F test statistical analysis used to determine the significance of the contribution of a single variable after all the other variables are in the equation.

Part 503 Rule Standards for the Use or Disposal of Sewage Sludge, 1993.

parts per billion (ppb) concentration expressed as the amount of a substance per a billion parts of a diluent.

parts per million (ppm) concentration expressed as the amount of a substance per a million parts of a diluent.

Parvoviridae a family of small (18–26 nm diameter), single-stranded DNA, nonenveloped icosahedral viruses.

parvovirus a genus of viruses in the Parvoviridae family; viruses have single-stranded DNA, are icosahedral in shape, and 18–26 nm in diameter; human parvoviruses have been associated with outbreaks gastroenteritis from shellfish consumption, rheumatoid arthritis, spontaneous abortion, and fetal death; majority of infections appear to be spread by the respiratory route, blood and blood products may also be involved in transmission.

passive fungal spore dispersal release of fungal spores without a discharge mechanism; generally associated with atmospheric disturbance such as wind currents or rainfall; in contrast to active fungal spore dispersal.

passive sampling collection of material without the use of a mechanical device; gravitational sampling.

Pasteur, Louis (1822–1895) chemist who studied problems of spoilage during fermentation caused by microorganisms; discoverer of lactic acid fermentation (1857), detailed the role of yeast in alcohol fermentation (1860), and settled the controversy of spontaneous generation (1864).

pasteurization process to eliminate the human pathogens and reduce the overall concentration of microorganisms in heat-sensitive foods; involves heating to approximately 68°C for 30 minutes then rapid cooling to 10°C or lower although industrial pasteurization uses higher temperatures for shorter periods of time.

pathogen/pathogenic an organism that infects a host and is capable of causing disease.

PBBs polybrominated biphenyls.

PBS phosphate buffer solution; phosphate-buffered saline.

PCA principal component analysis.

PCBs polychlorinated biphenyls.

PCP pentachlorophenol.

PCR polymerase chain reaction.

PCR-DGGE polymerase chain reaction—denaturing gradient gel electrophoresis.

PCRDU polymerase chain reaction detectable units.

Pearson product moment correlation coefficient (r) statistical value used with simple regression analysis that is equal to ± the square root of the coefficient of determination; a number between −1 and +1 with the sign the same as the slope of the regression line and the magnitude related to the degree of linear association between two variables.

pectinate comb-like in shape.

pedicel a slender stalk.

pedicellate having a short conidiophore.

pedoscope flattened glass capillary tube placed in soil to obtain a sample of microbial populations.

peer-review process in which scientific colleagues evaluate the findings of another investigator, generally used for reports, grant proposal and manuscript submissions.

PEG polyethylene glycol.

pelagic pertaining to the open ocean.

pellet concentrated deposition of organisms and other particulate material from a liquid suspension following centrifugation.

pellicle a dense firm mass formed on the surface of liquid growth medium.

PELs permissible exposure limits.

penicillate shaped like a small brush.

penicillinase enzyme added to liquid suspensions to minimize the antibacterial action of penicillin G.

penicillin G antibiotic used as an amendment to liquid and agar growth media to minimize the growth of gram-positive bacteria and some coccoid gram-negative bacteria.

Penicillium fungal genus; variegated blue, green or white colonies; loose spores cannot be distinguished from those of many species of *Aspergillus*; ubiquitous saprophyte isolated from soil, decaying plant debris, compost, foodstuffs especially fruit, cheese, fresh herbs, spices, cereals, nuts, and onions, and indoors on a variety of substrates including wallpaper, fabric, and paint; readily dispersed by wind and insects; industrial use in the production of cheese, sausages and antibiotics and antifungal agents; water activity of 0.78–0.86; associated with Type I allergies and Type III hypersensitivity; production of a variety of mycotoxins and microbial volatile organic compounds.

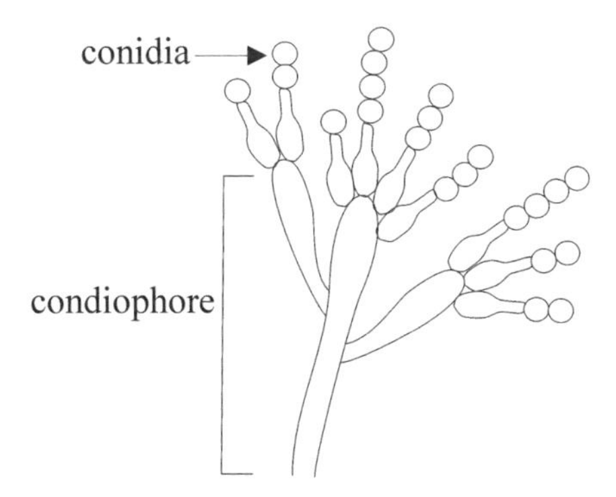

Penicillium camemberti fungal species; slow growing white to cream colony on malt extract agar with a cream yellow reverse; smooth or rough conidiophores with large, smooth subglobose conidia; isolated naturally from white fermented soft cheeses and on contaminated hard cheeses; associate with allergic reactions of cheese factory workers.

Penicillium chrysogenum fungal species; rapid growing velvet green colonies with a broad white margin and a yellow exudate with a yellow to cream reverse; smooth conidiophores with smooth, subglobose blue to dark green conidia; commonly isolated from dried foods and cereals, spices, desert sand, house dust; isolated from water damaged building, especially wetted wallpaper.

Penicillium notatum fungal species; former name for *Penicillium chrysogenum*; noted by Fleming in 1929 as producing penicillin, a substance inhibitory to a bacterial culture, leading to the discovery of antibiotics.

Penicillium roqueforti fungal species; broadly spreading velvet colonies that are green in color with a green-black reverse; roughly striped, large, smooth conidia; used in the production of blue-veined cheeses, but contaminated rye bread, corn silage, and homemade wine; mycotoxin production in silage and feed affecting cattle.

Penicillium verrucosum fungal species; production of ochratoxin in contaminated grain and meat products.

penicillus a complex of conidia-bearing branches that form a brush-like appearance.

penultimate next to the last.

peptide nucleic acid (PNA) probe nucleic acid sequence with a polyamide structure and individual nucleotide bases that is used to detect the presence of specific 16S rRNA in a sample.

peptidoglycan a polysaccharide peptide that forms a rigid layer that is responsible for the strength of the bacterial cell wall; a polymer of amino sugar chains cross-linked through tetrapeptide side chains; serves as a structural component of the gram-positive cell wall while in gram-negative bacteria it is covalently bound to a lipoprotein.

peptidyltransferase activity formation of peptide bonds catalyzed by ribosomal activity of the 23S rRNA.

peptone sorbitol bile salts pre-enrichment solution for the combined enrichment/polymerase chain reaction amplification detection of *Yersinia enterocolitica*.

percent recovery the amount of desired material retained following a treatment process.

perchlorate chlorine compound (ClO_4^-), used as the primary ingredient in solid rocket propellant, on the United States Environmental Protection Agency's drinking water contaminant candidate list due to its potential role in tumor formation and its presence in environmental waters.

perfect state sexual state of a fungus.

periphery the outer surface or boundary.

perithecia plural of perithecium.

perithecium a rounded, oval, or pyriform sac that actively dispenses asci by extruding or shooting them through an ostiole.

peritrichous flagella position of flagella at several locations around the bacterial cell; in contrast to polar flagella.

permissible exposure limits (PELs) concentrations established by the Occupational Safety and Health Administration as the amount or concentration of a hazardous substance in the air to which a worker can be exposed, based on an 8-hour time weighted average.

permissive host cell virus replication is accomplished within the host cell; in contrast to nonpermissive host cells.

persistent remains intact and/or infective.

person-to-person transmission dispersal of a microorganism from one individual to another.

Petri, Richard inventor of the petri plate for the culture of microorganisms in 1887.

petri plate a plastic or glass covered dish used for the culture of microorganisms.

Petroff-Hauser chamber hemocytometer.

PFA polyunsaturated fatty acid.

PFGE pulsed field gel electrophoresis.

PFU plaque-forming unit.

PGA poly(γ-D-glutamic acid).

pH negative log of the hydrogen ion activity; measurement of the level of acidity or alkalinity of a solution.

phage a short form of the word bacteriophage.

phagocytosis the ingestion of particulate material; contrast with pinocytosis.

PHAs polyhydroxyalkanoic acids.

Phallales fungal order; also known as stinkhorns that produce a foul odor and gelatinous ooze when releasing their basidiospores, resulting in the attraction of flies which transport the spores.

Phanerochaete chrysosporium fungal species; a member of the basidiomycetes; produces enzymes that digest lignin; a white rot fungus.

phase-contrast microscopy developed to improve contrast between specimens and the surrounding matrix for wet-mount preparations without the need for staining; a special ring in the objective lens increases the natural retardation of light in the matrix when the specimen and the matrix differ in refractive index resulting in the formation of a dark image on a light background; in contrast to dark-field microscopy.

phase separation the partitioning of two materials that are not longer miscible, often the result of a temperature change; used for the retrieval of desired material from background by partitioning between two liquids.

PHB poly-β-hydroxybutyrate.

phenol-chloroform extraction a method used to extract DNA or RNA from aqueous solution.

phenol/phenolic compound disinfectant for hard surfaces that acts by denaturing protein.

phenolic acid decarboxylases (PADs) enzymes present in some bacteria that are active in the conversion of phenolic acids to phenol derivatives which contribute to pleasant aroma in wines, but also can impart off-flavors in some food products.

phenotype the observable characteristics of an organism; in contrast to genotype.

phialides a bottle-shaped conidiogenous cell that produces conidia.

phialoconidium a conidium produced by a phialide.

phiX174/ϕX174 a single-stranded DNA, icosahedral coliphage; often used in environmental studies as an indicator for enteric viruses; a somatic phage, in contrast to a male-specific phage.

Phlebia radiata fungal species; a member of the basidiomycetes; a white rot fungus.

pH meter instrument used to measure the pH of a solution.

Phoma fungal genus; very small spores formed in pycnidia that are released through an ostiole; ubiquitous; approximately 80 species; isolated from plant material, soil, foodstuffs such as fruit, rice, and butter, and indoors on walls, ceiling tiles, flooring, cement, paint, paper, and wood; causative agent of black leg of crucifers; associated with Type I allergies and Type III hypersensitivity.

phosphate-buffered saline (PBS) a chemically-defined, balanced salt solution that is used as a diluent, a buffer, and in many cell culture applications.

phospholipid a polar lipid that is soluble in organic solvents with a hydrophilic group and an opposite end that is substituted with fatty acids, fatty aldehydes, or alcohols; associated with membranes.

phospholipid fatty acid (PLFA) analysis analytical method used to detect lipids in environmental samples for use in characterizing microbial populations; provides comparisons of total community lipid patterns and can detect shifts in specific microbial groups although many species have similar patterns.

phosphorus cycle the transformation of phosphorus in reservoirs by biological activity.

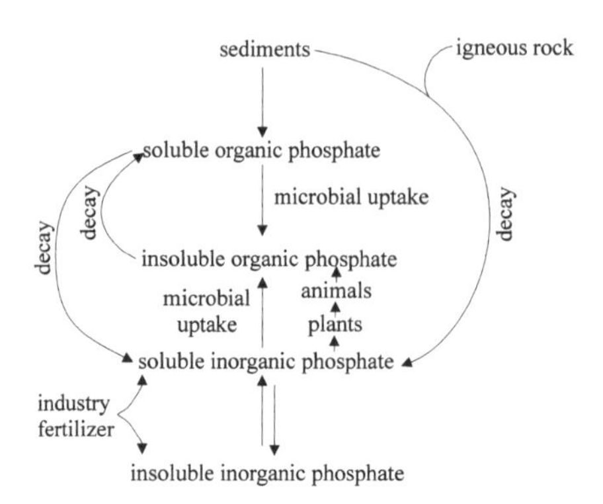

photoautotroph/photoautotrophic a microorganism that is capable of utilizing light as a sole energy source and CO_2 as a sole carbon source.

Photobacterium bacterial genus; member of the Vibrionaceae; ability to luminesce.

photochromogenesis formation of a pigment only when the microorganism is cultured in the light; used in the classification of some *Mycobacterium* spp.; in contrast to scotochromogenesis.

photochromogenic pigmented when exposed to light

photoheterotroph/photoheterotrophic organism that utilizes simple organic compounds as the carbon source and light as a sole energy source.

photometer instrument designed with a filter to generate a broad wavelength of light for the measurement of the turbidity of a microbial suspension; in contrast to a spectrophotometer.

photophosphorylation the synthesis of ATP using light as the energy source.

photoprotective agent material, pigment, or dye that serves as a shield against harmful exposure to light.

photoreactivation recovery of the activity of microorganisms in light that have been inactivated by ultraviolet light.

photo repair the repair of pyrimidine dimers in the DNA following damage due to exposure to ultraviolet light; involves the use of 310–480 nm near-UV light; in contrast to dark repair.

Photorhabus bacterial genus; gram-negative, rod-shaped symbionts of entomopathogenic nematode family Heterorhabiditidae which participate in maintaining suitable conditions for nematode reproduction; pathogenic to insects resulting from toxigenic reaction and septicemia; isolated from infected humans.

Photorhabus luminescens bacterial species; bioluminescent bacterium first isolated from a light-emitting insect.

photosynthesis/photosynthetic the conversion of light energy to chemical energy that involves light and dark reactions.

phototaxis movement toward light; observed with microorganisms as occurring either by scotophobotactic response or true movement toward the increasing intensity of a light source.

phototroph/phototrophic microorganism that is capable of using light as an energy source.

phototrophic bacteria distinguished from other bacteria by their ability to use light as their energy source; most are autotrophic while some are photoheterotrophic in their nutritional requirements; major groups include the cyanobacteria, green bacteria, *Halobacterium*, prochlorobacteria, and the purple bacteria.

phycobiont the primary producer population in a mutualistic relationship.

phyllosphere the surface of a plant leaf.

phylogenetic use of genetic traits to classify the relatedness of organisms.

phylogenetic probe a short base pair segment of DNA or RNA complementary to a gene that is representative of a primary phylogenetic domain, usually ribosomal genes.

phylogenetic tree illustration of the relative evolutionary positions of major groupings of organisms, also termed the universal tree of life.

phylogeny the history of the evolution of organisms.

physicochemical factors characteristics of an environment such as temperature, pH, oxygen, salinity and ionic strength.

phytanyl a 20-carbon branched chain in the lipid of Archaea.

phytopathogen bacterial plant pathogen; gram-negative representatives are certain species of *Erwinea*, *Pseudomonas*, *Xanothomonas*, and *Burkholderia*; gram-positive representatives are certain species of *Clavobacter*, *Corynebacterium* and *Streptomyces*; the specific disease caused is often diagnostic of the causative agent; endospores are not formed resulting in susceptibility to solar radiation and desiccation.

Phytophthora infestans fungal species; causative agent of potato blight.

Picornaviridae a family of nonenveloped, icosahedral viruses containing single-stranded RNA; human viruses in this family include the enterovirus and rhinovirus genera.

picornavirus a virus in the Picornaviridae family.

pileus the cap of a mushroom.

pili physical appendages on bacteria, similar in structure to fimbriae but are longer; generally only one or two are visualized on the bacterial cell when they serve as specific receptors of certain types of virus particles.

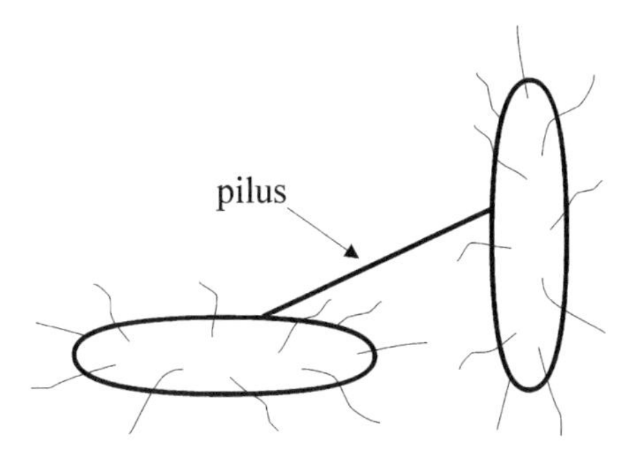

pilose covered with long, soft, hairy filaments.

pilot/pilot study an experimental design that is conducted on a small scale to provide relatively quick and inexpensive data that can be used to focus resources for a larger, more comprehensive study.

pilus singular term of pili.

pinocytosis uptake mechanism used by protozoa to engulf macromolecules in which the material is drawn into a channel formed by an invagination of the cell membrane that pinches closed resulting in the formation of a membrane-bound vacuole containing the macromolecule.

pioneer organisms the first microorganisms that colonize a habitat.

Pip a membrane protein that serves as a receptor protein for bacteriophage.

pipet/pipette small diameter cylinder, generally made of plastic or glass that is calibrated for the transfer of measured volumes of liquid.

pipetman/pipetteman a manually-powered or battery/electrically-powered device used with a disposable tip for the transfer of a specified amount of a liquid.

pip-shaped having the shape of an apple seed.

PLA poly(L-lactic acid).

plague rapid spread of infectious disease; the illness caused by the flea-borne transmission of *Yersinia pestis*.

planachromatic lens a flat-field achromatic lens that produces little curvature of field.

planapochromatic lens a flat-field apochromatic lens with minimal curvature of field.

Planctomycetales bacterial order; represented by obligate or facultative anaerobic chemoorganotrophic budding bacteria that have a protein cell wall without peptidoglycan and utilize *N*-acetylglucosamine as the sole carbon and nitrogen source; isolated from aquatic and upland soil habitats.

plankton very small, sometimes microscopic, plants and animals that are weakly swimming or drifting in a body of water.

planktonic relating to plankton.

plant infusion the use of plant material (e.g., rice or barley grain, boiled wheat) as a source of nutrients in growth media; commonly used in culture of flagellates and ciliates.

plaque a clear zone in a plaque assay indicating the presence of infective bacteriophage; also used to denote a thick film of *Streptococcus* spp. colonizing a saliva-originating glycoprotein layer on the surface of a tooth.

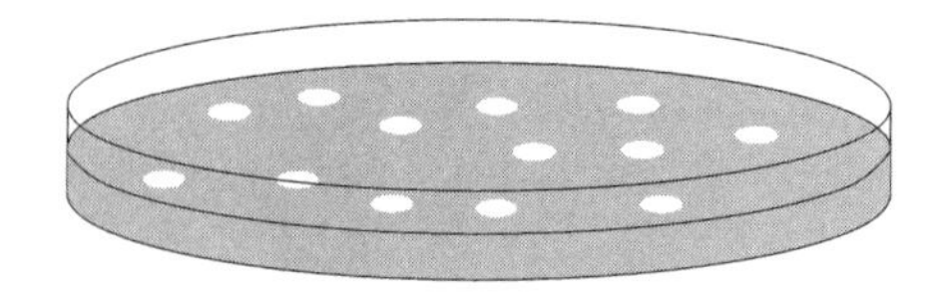

plaque assay culture-based method for the detection and enumeration of viruses; for bacteriophage, a serially diluted sample blended with a prepared suspension of host bacterial cells and a cooled, molten soft agar is poured onto a nutrient agar surface of a petri plate; for animal viruses, agar is poured over the surface of cells that have been exposed to a sample; in both cases, the samples are incubated for a prescribed period of time before zones of clearing (plaques) are enumerated; each clear zone indicates the lysis of the infected host cell by a virus particle.

plaque-forming unit (PFU) term used to refer to an infectious virus particle, or aggregate of virus particles, that, upon replication in cell culture, causes the formation of a single plaque.

plasmid a small, linear or circular piece of extrachromosomal cytoplasmic DNA that is capable of independent replication and can be transferred from one organism to another; may transfer genes (e.g., conferring antibiotic resistance); artificially constructed plasmids are used widely as cloning vectors in molecular biology.

plasmolysis the fragmentation of a cell membrane, often the result of changes in osmotic pressure and the loss of water.

plate count assay culture-based analysis in which the concentration of a microbial suspension is determined by applying an aliquot onto semisolid growth medium, incubating at suitable environmental conditions, and then enumerating colonies.

plate counter/plate reader instrument used to electronically enumerate the colonies isolated on the agar surface of a culture plate.

platykurtic reference to the kurtosis of a graph when it is more flat; in contrast to leptokurtic.

pleiotrophic multiple effects resulting from the action of a single gene or a mutation of a gene.

pleomorphic have more than one shape.

Pleurotus ostreatus fungal species; a white rot fungus; a ligninolytic organism that produces several laccase enzymes involved in the degradation of lignin, but it does not produce lignin peroxidase.

pleuston the interface region between the marine environment and the atmosphere with bacterial and algal inhabitants; the marine equivalent of the neuston.

PLFA phospholipid fatty acid.

plicate folded like a fan.

2-PLSM a microscopic technique that uses a high peak-power infrared laser with an extremely short pulse (femtosecond to picosecond range) to produce a high photon density resulting in the interaction of photons with a fluorescent dye to generate a signal.

pluriform having many different forms.

PM-2.5 designation by the United States Environmental Protection Agency for airborne particulate matter that is <2.5 μm in diameter; associated with industrial and residential combustion and vehicle exhaust; also termed fine particulate.

PM-10 designation by the United States Environmental Protection Agency for particulate matter that is <10 μm in diameter and considered a criteria air pollutant; includes dust, soot and other airborne particulate generated or released into the outdoor air; associated with eye, nose and throat irritation, and other respiratory illnesses.

PNA probe peptide nucleic acid probe.

pneumoconiosis disease characterized by the permanent deposition of particulate in the lungs.

POC particulate organic carbon.

point graph dot plot.

Poisson distribution the distribution of random events occurring in a defined period of time, and with the mean equal to the variance.

polar water soluble; hydrophilic.

polar flagella position of flagella at one or both ends of the bacterial cell; in contrast to peritrichous flagella.

poliovirus an enterovirus, member of the Picornaviridae family, causative agent of paralytic poliomyelitis, and aseptic meningitis; shed in feces of infected individuals and transmitted by the fecal-oral route; isolated from drinking water, ground water, and wastewater; used as a model organism for enteric viruses.

polyaromatic hydrocarbons (PAHs) organic compounds consisting of multiple rings.

polyacrylamide gel electrophoresis (PAGE) an electrophoresis technique for the separation of molecules based on their electrophoretic mobility in a gel matrix.

polybrominated biphenyls (PBBs) chemical compounds used as flame retardants that consist of biphenyls with bromine substitutions resulting in resistance to biodegradation.

polycentric having many centers of growth and differentiation.

polychlorinated biphenyls (PCBs) chemical compounds used as plasticizers in polymers in electrical condensers and transformers, insulation materials, batteries, flame retardants, lubricants, and heat exchange fluids that consist of biphenyls with 1—10 chlorine atoms per molecule resulting in increased resistance to biodegradation with increased numbers of chlorine substitutions; toxic to humans.

polyclonal antibody an antibody produced by several clones of B lymphocytes; usually prepared in rabbits; in contrast to monoclonal antibody.

polyethylene glycol (PEG) a clear, colorless, viscous liquid that is soluble in water; used in pharmaceuticals, food products, and as a binder in industry.

poly(γ-D-glutamic acid) (PGA) a polyamino acid, water-soluble slime capsular substance that is produced by a variety of bacteria; substance that can sequester ammonia for slow microbial degradation in soils.

polyhedric a many sided, three-dimensional figure.

polyhydroxyalkanoic acids (PHAs) compound synthesized by several bacterial species when there is an abundance of carbon present compared to nitrogen or phosphorus; shown to increase bacterial survival in nutrient-depleted situations.

poly-β-hydroxybutyrate (PHB) storage material present as inclusion bodies in prokaryotic organisms.

poly(L-lactic acid)/polylactide (PLA) a renewable, biodegradable plastic that is produced by chemical synthesis or the fermentation of starchy materials, cane molasses, and cellulose; degraded by microorganisms during composting.

polymer a macromolecule consisting of a repeating structural unit.

polymerase an enzyme that uses a template to cause the synthesis of RNA or DNA from nucleotide bases.

polymerase chain reaction (PCR) amplification molecular biology technique in which the concentration of a unique gene sequence is exponentially increased by a series of temperature-dependent reactions.

polymerase chain reaction detectable unit (PCRDU) term used to refer to a single copy of a genome that is detectable by polymerase chain reaction(PCR) amplification; contrast with colony-forming unit and plaque-forming unit.

polymerization the formation of a polymer; extension of a gene sequence during the polymerase chain reaction (PCR) amplification process.

polymorphic having many forms.

polyomavirus a genus of viruses in the family Papovaviridae; cause a variety of tumors in humans.

polypodial having more than one pseudopod; used to describe amoeba; in contrast to monopodial.

polypore descriptive of a wood-inhabiting basidiomycete; associated with wood decay and buildings constructed of wood.

polyunsaturated fatty acid (PFA) recently recognized as 35–45 carbon chains that are distinct from that found in plants and animals; formerly unknown in bacteria.

Pontiac fever human respiratory disease characterized as an influenza-like illness resulting from inhalation exposure to legionellae; mean incubation period of 36 hours with a greater than 95% attack rate; virulence and viability of the contaminant bacterium are suggested as reasons for this illness in contrast to Legionnaires' disease; first documented in Pontiac, Michigan.

population-based prevention distribution of control measures to an entire population; in contrast to risk-based prevention.

porins channels in the lipopolysaccharide layer of gram-negative bacteria for the transport of small molecules.

poroconidia fungal conidia formed when the inner wall of a conidiogenous cell pushes through a pore in the outer wall; observed with *Alternaria* and *Ulocladium*.

porosity the ratio of open space to the bulk volume of a material; the number of interstitial spaces in a soil or sediment mixture.

porters membrane-bound proteins that function in the transport of substances in and out of a cell.

positive feedback mechanism in which the formation of a product increases the production of product; in contrast to negative feedback.

positive hole correction mathematical correction provided by manufacturers of sieve-type impactor air samplers (e.g., Andersen single-stage air sampler) to correct for the collection of multiple fungal spores or bacterial cells through the same hole; failure to correct the data will result in underestimation of the airborne concentration.

positive predictive value the ability of a test method to correctly identify true positives; calculated by dividing the number of true positives by the total number of positives, a value that includes both the true positives and the false positives.

positive skew the shifting of the symmetry of a curve to the right; see skew.

potable drinkable.

potato dextrose agar (PDA)/potato glucose agar a fungal culture medium that favors sporulation, consists of potatoes, agar, dextrose, and water.

potato flake agar fungal culture medium that favors sporulation.

potato glucose agar (PDA) a fungal culture medium that favors sporulation.

pounds per square inch (PSI) a measure of pressure that is equal to the weight of one pound on a square inch of area.

pour plate culture and isolation method in which a measured volume of a microbial suspension is blended with cooled, unsolidified agar growth medium prior to dispensing into a petri plate; resulting colonies are formed within the agar, not on the surface as observed with spread plate culture methods.

powdery mildews diseases of plants caused by fungi, notably *Oidium*, that are characterized by a light colored powdery dusting on leaves and stems of affected plants.

power of the test the probability of concluding that there was a difference when there was a difference.

ppb parts per billion.

ppm parts per million.

PRD1 a double-stranded DNA bacteriophage containing lipid in its capsid; member of the tectivirus genus in the family Tectiviridae; often used in studies of virus survival and transport in the environment.

precipitation a reaction of two or more compounds resulting in visible accumulation of product.

precision the reproducibility of a measurement; in contrast to accuracy.

predation parasitism in which one organism engulfs another.

predictive value the ability of a test method to correctly identify the presence or absence of an analyte of interest; generally classified as positive predictive value and negative predictive value.

preemptive colonization a situation in which establishment by pioneer organisms alters the conditions of the habitat, thereby preventing further succession.

prefilter matrix of fibrous material used to retain larger particles prior to additional filtration procedure(s).

presence-absence (P-A) test term used to describe any one of several analytical methods that detect the presence or absence of microorganisms, rather than quantifying the number of organisms present in a sample.

presumptive water quality testing evaluation of water samples for total coliform bacteria using lauryl-tryptose broth-supplied test tubes with a Durham tube; results are recorded as positive for gas production at 35°C for 24 hours indicative of the presence of coliform bacteria, although certain noncoliform bacteria may also produce these results, thereby requiring confirmatory water quality testing.

prevalence the number of individuals having a specific characteristic (e.g., disease) present in a population at a specific time divided by the total number of individuals in that population; in contrast to incidence.

Prevotella bacterial genus; anaerobic, gram-negative, straight, helical or curved, oligosaccharolytic and xylanolytic bacilli that are isolated from the rumen.

Prevotella brevis bacterial species; one of the complex of microorganisms isolated from the rumen; formerly termed *Bacteroides ruminicola* subspecies *brevis*.

Prevotella bryantii bacterial species; one of the complex of microorganisms isolated from the rumen that is active in the degradation of starch.

Prevotella ruminicola bacterial species; one of the complex of microorganisms isolated from the rumen that is dominant when the host animal is fed a hay diet.

primary cell culture a culture prepared directly from an organ that has been treated to disperse the cells into a single-cell suspension that is then added to a flask and allowed to form a monolayer; in contrast to a continuous cell line.

primary metabolite material produced during the growth phase of the microbial life cycle as a product or by-product of metabolism.

primary phylogenetic domain grouping of organisms at the most general level.

primary prevention an action taken to prevent the development of disease in a well population; in contrast to secondary prevention.

primary producer/primary production autotrophic metabolism resulting from the synthesis of organic matter from inorganic sources.

primary stain first dye applied to a specimen that is subsequently decolorized and treated with a counterstain.

primary standard a regulatory pollution limit value for a criteria air pollutant or drinking water contaminant that is based on health effects; in contrast to a secondary standard.

primary succession the establishment of the first population of organisms in a habitat.

primary wastewater treatment physical separation of sewage materials as the first step in the treatment process.

primer a polynucleotide molecule that serves as the attachment site for the first nucleotide during DNA replication.

principal component analysis (PCA) statistical analysis in which the structure of a relationship is examined without distinctions being made between the dependent variables and the independent variables.

prion an infectious agent that does not contain nucleic acid.

probabilistic model term used to describe a mathematical model in which the input variables, and consequently the output information, are described by a probability distribution of values, rather than a single value; in contrast to deterministic model.

probability the relative likelihood that an event will or will not occur relative to other events.

probe a defined sequence of single-stranded nucleic acids that is used to detect the presence of complementary nucleic acid sequences in a sample.

probiotic living microorganism in a food product that provides an additional health benefit beyond inherent basic nutrition.

prochlorobacteria a group of phototrophic bacteria consisting of a single recognized genus (*Prochloron*) that use O_2 as their carbon source and water as their electron donor for oxygenic photosynthesis.

Prochloron bacterial genus; the only recognized genus of prochlorobacteria; single—celled extracellular bacteria that are symbionts of marine invertebrates.

prochlorophyte an oxygenic phototroph that contains chlorophyll a and b, but lack phycobilins.

procins bacteriocins produced by *Pseudomonas* species.

profundal zone the section of freshwater where sunlight penetrates; inhabited by an abundance of microorganisms with the majority of the population being secondary producers.

prognostic selection length-based sampling.

prokaryote/prokaryotic a microorganism generally <2 µm in size with a single DNA molecule that is usually circular and does not have a membrane-bound nucleus; in contrast to eukaryote.

prokaryotic flagellum flagellum.

proliferation an increase in abundance.

proline an osmolyte that may serve as a virulence factor for some bacteria.

propachlor *N*-isopropyl-2-chloracetanilide, a pesticide with a secondary amine group that is not cleaved by microbial amidase enzymes and remains in the environment; in contrast to propham.

propagation the reproduction of an organism.

propagule a single cell or single reproductive unit that is capable of reproduction.

prophage formed when the genome from a lysogenic bacteriophage is integrated into the host bacterium's chromosome; it is replicated with the host's DNA.

propham isopropyl-*N*-phenylcarbamate, a pesticide with a primary amine group that is rapidly cleaved by microbial amidase enzymes and does not persist in nature; in contrast to propachlor.

propidium iodide a fluorescent stain used for the detection of DNA and nonviable cells.

Propionibacterium bacterial genus; anaerobic to aerotolerant, pleomorphic, gram-positive, non-motile, chemoorganotrophic bacilli; ferment glucose to propionic acid or acetic acid.

Propionibacterium freundenreichii bacterial species; a gram-positive, non-spore-forming, facultatively anaerobic, nonmotile propionic acid bacterium that is found in dairy products and on the skin; used in the manufacture of Swiss-type cheese and the industrial production of propionic acid and vitamin B_{12}.

propionic acid bacterium (PAB) general term used to describe a gram-positive, nonmotile, non-spore-forming, facultatively anaerobic organism.

prospective study the selection of two populations, one exposed and other nonexposed, to determine if the exposure results in an observable change over time.

prosthecae a budding or an appendage that extends or protrudes from the cell of budding bacteria.

protein coat the structural component of a virus particle that encloses the nucleic acid; also called a capsid.

Proteobacteria largest division of bacteria that includes the chemoorganotrophs, phototrophs, and chemolithotrophs.

proteolytic capable of degrading proteins.

Protista one of the kingdoms in the five-kingdom classification of biology that includes the unicellular eukaryotic microorganisms, including protozoa, unicellular algae, and Myxomycetes.

protobiotic preparation of microorganisms or microbial products that demonstrate a beneficial effect for the host.

protocol details of the method(s) used to obtain data.

protocooperation synergism.

protomer the proteins in a virus capsid that form clusters called capsomeres.

proton motive force the energized state in a membrane that is created by the expulsion of protons.

prototroph the parent organism from which an auxotroph is produced.

protozoa unicellular eukaryotic microorganisms that lack cell walls, feed by ingestion of particulate or macromolecular materials and are generally classified into groups based on their means of motility; unlike algae they lack chlorophyll and unlike slime molds they do not have fruiting structures.

protozoonotic an organism that naturally infects protozoa.

provirus prophage.

proximal near the central portion of the body.

pruinose powdery.

pseudo false.

pseudohyphae elongated cells or cells that remain attached to each other as a chain.

Pseudomonaceae bacterial family; members are characterized as ubiquitous gram-negative bacilli that are nonsporulating and versatile in their nutrient requirements.

Pseudomonas bacterial genus, member of the family Pseudomonaceae; typically straight bacilli that are motile by polar flagella, although one species is nonmotile; mesophilic but some species grow at temperatures between 4°C and 41°C; many species can utilize nitrate as a nitrogen source and some carry out denitrification; antibiotic fluorescent species used as biocontrol agents.

Pseudomonas aeruginosa bacterial species; ubiquitous soil organism that proliferates in standing water and produces a characteristic grape-like odor; opportunistic pathogen especially in immunocompromised patients.

Pseudomonas angulata bacterial species; phytopathogen, causative agent of leaf spot of tobacco.

***Pseudomonas* CAM-1** bacterial species; a psychrotolerant organism that degrades polychlorinated biphenyl (PCB).

Pseudomonas carboxidoflava bacterial species; aerobically utilizes carbon monoxide as a source of carbon and energy.

Pseudomonas carboxidohydrogena bacterial species; aerobically utilizes carbon monoxide as a source of carbon and energy.

Pseudomonas caryophylli bacterial species; phytopathogen, causative agent of wilt of carnation.

Pseudomonas cepacia *Burkholderia cepacia*.

Pseudomonas fluorescens A506 bacterial strain; a naturally occurring ice$^-$ strain licensed for the control of frost injury to the pear.

***Pseudomonas fluorescens* CHA0** bacterial strain; a naturally occurring biocontrol agent that suppresses black rot of tobacco.

Pseudomonas fluorescens DR54 bacterial strain; an isolate from the sugar beet rhizosphere that is effective as a biocontrol agent by preventing pre-emergent damping-off disease caused by fungal pathogens.

Pseudomonas glycinea bacterial species; phytopathogen causative agent of blight of soybeans.

Pseudomonas marginalis bacterial species; soft-rot phytopathogen resulting in infection of plant stems and shoots, but not leaves and causative agent of slippery skin of onion.

Pseudomonas marginata bacterial species; phytopathogen, causative agent of scab of gladiolus.

Pseudomonas phaseolicola bacterial species; phytopathogen, causative agent of halo blight of beans.

Pseudomonas pisi bacterial species; phytopathogen, causative agent of blight of peas.

Pseudomonas putida bacterial species; colonize root surfaces and the rhizosphere; many strains degrade toxic compounds in the soil; some strains promote plant growth and some are useful as biocontrol agents.

Pseudomonas syringae bacterial species; common leaf bacterium but also a phytopathogen resulting in chlorotic lesions of leaves, and the causative agent of blight of lilac; isolates have been genetically modified to delete a gene sequence that controls production of a cell wall protein rendering the organism unable to serve as an ice nucleation site (ice^-) thereby preventing frost damage to crops.

Pseudomonas tabaci bacterial species; phytopathogen, causative agent of wildfire of tobacco.

pseudopod a blunt ended projection from a cell; mechanism of locomotion for amoeba.

PSI/psi pounds per square inch.

psychrometer type of hygrometer that uses two similar thermometers, one a wet bulb and the other a dry bulb, the difference in the readings is a measure of the dryness of the atmosphere.

psychrophile/psychrophilic descriptive of the temperature requirements of microorganisms with an optimal growth temperature less than 20°C.

psychrotolerant microorganisms that are able to grow at low temperature but have an optimal growth temperature >15°C.

public health the health of a population.

Puccinia fungal genus; member of the Basidiomycotina; causative agent of rust of cereals.

puffball description of the passive release of the cloud of spores released by basidiospores when impacted by rain drops, humans, or small animals.

Pullularia pullans fungal genus; former terminology for *Aureobasidium pullulans*.

pulmonary relating to or associated with the lungs.

pulsed field gel electrophoresis (PFGE) an electrophoresis technique that permits the separation of larger molecules (>40,000 base pairs) using short pulses of electricity into an array of electrodes that surround an agarose gel matrix.

punctate having points.

pure culture the presence of a single microbial species in suspension or on a culture surface.

purine one of the molecules that comprises nucleic acids; these include adenine and guanine.

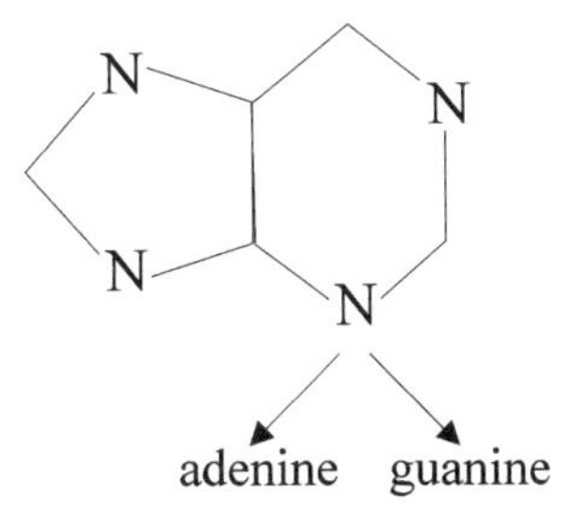

purple bacteria a heterogeneous grouping of anoxygenic phototrophic bacteria that conduct photophosphorylation using Bchl *a* and Bchl *b* chlorophyll pigments with organic carbon or CO_2 as their carbon sources and H_2, H_2S, or sulfur as their energy source; in contrast to green bacteria.

***P* value** a statistical term; small *P* values signify that the null hypothesis should be rejected.

pycnidia plural of pycnidium.

pycnidium asexual fungal fruiting structure; represented in the genus *Phoma*.

pyogenic pus-producing.

pyriform pear-shaped.

pyrimidine one of the molecules that comprises nucleic acids; these include cytosine, thymine, and uracil.

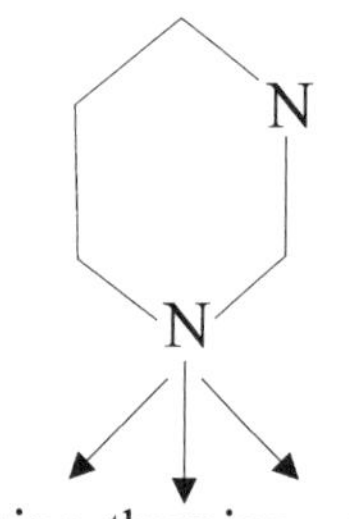

pyrimidine dimer a structure that results from the ultraviolet irradiation of DNA in which two adjacent pyrimidine bases become covalently linked; the presence of the dimers prohibits DNA replication.

Pyrobaculum bacterial genus; a member of the Archaea that is unique in their aerobic respiration and denitrification using either organic or inorganic electron donors; thermophilic organism that grows at temperatures up to 103°C; growth is inhibited by the presence of S°.

Pyrococcus bacterial genus; member of the Archaea; hyperthermophilic cocci that grow at a temperature range of 70–106°C with an optimal growth temperature of 100°C and uses proteins, maltose and starch as electron acceptors in the reduction of S° to H_2S.

Pyrococcus abyssi bacterial species; a heterotrophic hyperthermophilic euryarchaeon isolated from deep-sea thermal vents.

Pyrodictium bacterial genus; member of the Archaea; anaerobic, chemolithotrophic, submarine, volcanic hyperthermophilic organisms with a temperature range of 82–110°C and an optimal growth temperature of 105°C; the bacteria produce a mass consisting of hollow fibers that attach the cells to a solid surface.

pyrogenic fever-producing.

Pyrolobus bacterial genus; member of the Archaea; submarine hyperthermophilic organisms that are capable of growth up to 113°C and reduce NO_3^- to NH_4^+ using H_2 as the electron donor.

Q

QA quality assurance.

QAP quality assurance plan.

Qbeta, Qβ a male-specific, single-stranded RNA bacteriophage in the Leviviridae family.

QC quality control.

QMP quality management plan.

QMRA quantitative microbial risk assessment.

QPCR quantitative polymerase chain reaction.

Qualicon RiboPrinter commercially available system for ribotyping that lyses cells in a bacterial suspension, extracts the DNA, digests the DNA with a restriction endonuclease, separates the digests on a gel, transfers the DNA bands to a membrane, probes the bands with non-radioisotope-labeled rRNA-specific probes, photographs the membrane, and then compares the bar code-like pattern to databases for identification of the genus and species.

quality assurance (QA) a management philosophy to assure the reliability of data; the integration of planning, assessment, and interpretation of study data.

quality assurance plan (QAP) a document that specifies the quality assurance and quality control requirements for each project being conducted.

quality control (QC) the routine application of procedures to obtain predetermined standards of performance.

quality management plan (QMP) a document that specifies the policies for a study directed from the corporate or management level to the laboratory and field scientists.

quantitative microbial risk assessment (QMRA) a scientific process used to calculate the probability that a specified outcome (e.g., infection, morbidity, or mortality) will occur as a result of exposure to a specified dose of microorganisms over a specified period of time.

quantitative polymerase chain reaction (QPCR) amplification molecular biology technique in which the concentration of a unique gene sequence is enumerated following an exponential increase produced by polymerase chain reaction amplification.

quaternary ammonium compound a cationic detergent that is used to disinfect hard surfaces by interacting with the phospholipids of microbial membranes.

questionnaire a listing of questions presented to a population to determine if there is a commonality among the individuals; often used in field investigations to determine if there has been an exposure or concern for an exposure.

quorum-sensing (QS) a density-dependent phenomenon in which a reaction occurs when a substance accumulates to a sufficient concentration; described in gram-negative and gram-positive bacteria and *Streptomyces*; associated with the bioluminescence of *Vibrio fischeri*.

QS quorum-sensing.

Q_{10} value the change in enzymatic activity caused by a 10°C temperature rise; also termed the ten-degree temperature quotient.

R

r Pearson product moment correlation coefficient.

R multiple correlation coefficient.

R^2 coefficient of determination.

R2A agar a low nutrient culture medium used for the isolation of oligotrophic bacteria, primarily for isolation of bacteria in water samples.

RAB rotating annular bioreactor.

racemes/racemic a mixture of substances that does not rotate plane polarized light.

racquet hyphae fungal hyphae that are composed of cells that are inflated at one end, resembling the shape of a tennis racquet or a snowshoe.

radiate a wheel-shape formation around an axis.

radioimmunoassay (RIA) immunoassay method that uses radioisotopes to increase the sensitivity of antibody detection.

raft/rafting the use of a larger particulate or fragment by a smaller cell to increase airborne or waterborne transport.

Rainbow Agar 0157 a commercially available, selective chromogenic agar for the detection of *E. coli* O157:H7 which grows as black colonies.

Raman spectroscopy an analytical technique used for the characterization of materials with similar crystalline structure or stoichiometry.

ramp variation a method for refuse disposal in a sanitary landfill in which solid waste is spread and compacted on a slope with daily coverage of soil taken from the base of the ramp; in contrast to the trench method or the area method.

random factor statistical term to denote that the specific samples tested in an assay are only representative examples of a larger group of possible choices; in contrast to a fixed factor.

randomized trial an experimental study in which the test and control populations are selected without a pattern or preselected scheme; in contrast to a stratified randomized trial.

randomly amplified polymorphic DNA (RAPD) a method in molecular biology in which small primers attach to DNA in a pattern not linked to specific genes; used in polymerase chain reaction amplification to create an array of DNA fragments (i.e., a DNA fingerprint); the fingerprints from two different organisms can then be compared and the degree of relatedness of the organisms can be determined; the method is used when the sequence of the target nucleic acid is unknown or little is known about the sequence.

random-number table/random-number generator use of a printed format available in a statistical manual or computer-based program for the selection of test and control samples in a randomized trial.

range measurement of dispersion; the difference between the highest and lowest values; presented as a single value, not as a spread of numbers.

RAPD randomly amplified polymorphic DNA.

rarefaction a statistical approach that compares observed richness among sites that have not been sampled equally.

ratio variable data existing in an ordered category where there are equal differences between values and the zero point is meaningful; descriptive of most laboratory data, in contrast to an interval variable.

RCS Plus sampler a battery-powered, impactor sampler used for the collection of bioaerosols at a fixed flow rate of 50 liter/min. onto small agar-filled wells located on a plastic strip for follow-on culture analysis.

rDNA ribosomal DNA.

ready-to-eat foods (RTE) food or food products, other than fruits and vegetables, that are eaten without heating by the consumer.

reannealing the reformation of a DNA double strand that was disassociated by heating.

recalcitrant totally resistant to biodegradation.

reclaimed water treated domestic wastewater effluent intended for beneficial reuse.

recombinant DNA a segment of DNA that contains nucleic acid from two or more sources.

recombination the combination of genetic elements from two different sources into a single unit.

recreational water the water in a natural (e.g., lakes, rivers) or artificial setting (e.g., swimming pools, water parks) that is used for recreational purposes (e.g., swimming, water skiing, canoeing); current ambient water quality criteria for recreational water established by the United States Enivironmental Protection Agency are: for fresh water, enterococci: 33 per 100 ml or *E. coli* 126 per 100 ml (geometric mean) and for marine water, enterococci 35 per 100 ml (geometric mean).

recurve bend backwards.

recycled water reclaimed water.

redox potential (E_h) the proportion of oxidized components relative to reduced components with a positive value favoring oxidation and a negative value indicating a reducing environment; expressed in mV.

redox reaction the oxidation of one compound paired with the reduction of another compound.

red tide coloration of marine waters by dinoflagellates that generally covers several square kilometers and produce compounds that are toxic to fish and other marine organisms.

reducing agent substance that combines with oxygen or loses electrons in a reaction.

reduction a chemical or biochemical reaction in which the reactant gains electrons.

reductive dechlorination a reductive dehalogenation in which there is a release of Cl^- during an aerobic respiration with a chlorinated organic compound serving as an electron acceptor.

reductive dehalogenation the release of any halogen (e.g., Cl^-, Br) during anaerobic respiration with the halogen serving as an electron acceptor.

referral bias error introduced in an epidemiological survey resulting from the selection of individuals with specific characteristics; individuals are attracted to participate in a study at a disproportionate rate because of certain characteristics and the nature of the study.

refractive bending of light.

refractive index a value given to the change in the velocity of light as it penetrates a substance resulting in the bending of the path of the light.

refrigeration storage the placement of materials at 2–6°C.

refringent refractive.

regolith rock rubble.

regression simple regression or multiple regression.

regression line a straight line passing through a plot of simple regression data that minimizes the sum of the squared differences between the original data and the fitted points.

relative humidity (RH) the ratio of the actual vapor pressure of air to the saturation vapor pressure; listed as a percentage.

relative risk an interpretation of exposure data to determine the etiologic relationship between exposure and disease; calculated as the risk in the exposed population divided by the risk in the nonexposed population, with a value greater than one indicative of a positive association between exposure and illness and a value less than one indicative of a negative association.

reliability repeatability.

remediation the removal and decontamination of a microbial-contaminated area.

reniform kidney-shaped.

Reoviridae a family of icosahedral, double-stranded RNA viruses, 60–80 nm in diameter; includes the reovirus, rotavirus, and orbivirus families of human pathogenic viruses.

reovirus a genus of icosahedral double-stranded RNA viruses, 80 nm in diameter, in the Reoviridae family; cause gastrointestinal and respiratory illnesses; transmitted by the fecal-oral route.

repeatability the ability of a method to replicate results previously obtained with the same protocol; also termed reliability.

repeated-measures analysis of variance/repeated-measures ANOVA statistical analysis used when repeated observations are recorded on each test subject.

repetitive sequence-based polymerase chain reaction amplification (REP-PCR) a polymerase chain reaction amplification technique that is used as a rapid fingerprinting method at the strain level of identification.

replica plating transfer of microorganisms in culture from one agar surface to another using a sterile velvet cloth or filter matrix.

replicate organism detection and counting (RODAC) plate 65 × 15 mm diameter plastic culture dish that is filled with an agar medium to form a convex surface for sampling of smooth surfaces to determine the presence of culturable bacterial or fungal contamination.

replication the process of making a copy, for example of a molecule such as DNA or RNA.

REP-PCR repetitive sequence-based polymerase chain reaction amplification.

resazurin a redox-indicating dye added to thioglycolate broth to demonstrate the depth of oxygen penetration in the liquid.

reservoir the environment where a microorganism is established; a large body of surface water used for recreation and/or as a source of drinking water.

resolution the ability to distinguish two adjacent objects; this ability is not limitless and it is dictated by the physical properties of light.

resolving power the measured resolution of a microscope; this is a function of the wavelength of light used and the numerical aperture of the objective lens.

respirator breathing apparatus designed to filter out biological and/or chemical agents depending on the filtration matrix used.

respirator fit test validation of the wearing of a respirator to ensure proper fit.

respiratory route of exposure inhalation route of exposure.

respiratory system the breathing system of humans including the mouth, nose, larynx, trachea, and lungs with associated nerves and blood supply; often illustrated as being divided into the upper respiratory tract and the lower respiratory system.

restriction endonuclease an enzyme that recognizes a specific sequence of nucleotides in a DNA molecule, then catalyzes the cleavage of the bonds between specific nucleotides within that sequence.

restriction enzyme restriction endonuclease.

restriction fragment length polymorphisms (RFLP) differences in the nucleotide base sequences (such as those that would be produced by a mutation, or a segment of DNA that contains a different number of repeat sequences) at the restriction sites of different individuals within a species that result in fragments of different lengths when treated with restriction endonucleases; used to discriminate between closely related individuals.

resuspension the return of settled particulate into suspension, such as would be produced by shaking.

reticulate netted.

retrospective cohort study use of historical data to shorten the time of an epidemiology study of a select population; in contrast to a longitudinal study.

retrovirus any virus in the family Retroviridae; viruses are icosahedral in shape, surrounded by a lipid envelope, contain single-stranded RNA; human pathogens in this family cause different types of cancer, and the human immunodeficiency viruses (causative agent of AIDS) are members of the lentivirus genus in this family.

reverse osmosis a water treatment process in which a membrane is used to reduce the concentration of salt and other contaminants from water; water moves in an opposite direction to the concentration gradient, from a high salt environment to a lower saline environment, thus pressure is required.

reverse transcription the copying of information from RNA to DNA.

reverse transcription polymerase chain reaction (RT-PCR) an amplification process in which RNA is used as a template for the synthesis of cDNA, the cDNA is then amplified using polymerase chain reaction amplification.

RFLP restriction fragment length polymorphisms.

RH relative humidity.

rhicadhesin a calcium binding adhesion protein on the surface of *Rhizobium* and *Bradyrhizobium*.

rhinovirus a genus of viruses in the family Picornaviridae; the most commonly isolated viruses in persons with the common cold.

rhizobia common name for members of the Rhizobiaceae.

Rhizobiaceae bacterial family; characterized by their ability to fix atmospheric nitrogen.

Rhizobium bacterial genus; gram-negative, aerobic, chemoorganotrophic, pleomorphic bacilli that are motile by one polar flagellum or by 2–6 peritrichous flagella; present in root nodules and involved in nitrogen fixation.

rhizoid root-like structure of branched filaments used by some fungi to obtain nutrients.

Rhizopus fungal genus; fast-growing colony with aerial hyphae, stolons, pigmented rhizoids, sporangiophores and sporangia; species may be phytopathogens or animal parasites.

rhizosheath relatively thick cylinder of soil that adheres to plant roots; typical of desert grasses and found in other grasses with sand grains cemented together by an extracellular mucigel excreted by the root cells.

rhizosphere region immediately surrounding a plant root; thin layer of soil that adheres to a root system after the loose soil has been removed.

rhodamine B an organic dye that fluoresces red; used for the detection of microbial cells.

Rhodobacter bacterial genus; gram-negative ovoid or rod-shaped cells that may produce a slime or capsule; photoautotrophic in the presence of sulfide and photoheterotrophic under lighted anaerobic conditions; widely distributed in freshwater, marine, and hypersaline environments.

Rhodobacter sphaeroides bacterial species; photosynthetic bacterium that sequesters metal such as tellurite and efficiently reduces selenite with intracellular accumulation under both dark and anaerobic photosynthetic conditions.

Rhodococcus bacterial genus; aerobic, chemoorganotrophic, gram-positive, cocci to short rods that form filaments with side projections; widely distributed in nature and abundant in soil and dung of herbivores; some species are pathogenic to humans and animals.

Rhodococcus erythropolis bacterial species; demonstrates microbial desulfurization with the ability to specifically remove the sulfur atom from dibenzothiophene.

Rhodospirillum bacterial genus; spiral cells that are motile by polar flagella with internal photosynthetic membranes in vacuoles or lamellae; photoheterotrophic under lighted anaerobic conditions; isolated from freshwater and marine environments.

Rhodospirillum rubrum bacterial species; conducts extracellular reduction of selenite.

Rhodosporidium fungal genus; basidiomycetous yeast found in marine environments.

Rhodotorula fungal genus; isolated from moist areas on the skin and from wetted environmental surfaces such as grout and shower curtains; cells are oval-shaped, multilateral budding yeasts with a capsule.

RIA radioimmunoasssay.

ribonuclease (RNase) an enzyme that catalyzes the cleavage of RNA.

ribonucleic acid (RNA) a macromolecule composed of the sugar ribose and the purine bases adenine and guanine and the pyrimidine bases cytosine and uracil (rather than thymine as in DNA); present in cells as messenger RNA, transfer RNA, and ribosomal RNA.

ribosomal DNA (rDNA) DNA that codes for ribosomal RNA (rRNA).

ribosomal RNA (rRNA) the RNA portion of a ribosome.

ribosome one of the components involved in the synthesis of proteins from DNA; composed of rRNA and protein; ribosomes are composed of two subunits, a smaller unit and a larger unit; in prokaryotes, the small unit contains 16S rRNA and the large subunit contains 5S rRNA and 23S rRNA; in eukaryotes, the small unit contains 18S rRNA (16S rRNA in plants) and the large subunit contains 5S rRNA, 5.8S rRNA, and 28S rRNA.

ribotyping molecular biology methodology that scores restriction polymorphisms in the rRNA operons of prokaryotes; used to identify similarities in rRNA gene sequences for genotypic determination of isolates.

rice medium a fungal culture medium.

richness species richness.

richness estimators a statistical approach in which an estimate of the total richness of a community is made from a sample and then the estimates are compared across samples.

Ricketts, Howard scientist who discovered rickettsia.

Rickettsia bacterial genus; short, gram-negative, nonmotile bacilli that have not been isolated on artificial media, but require host cells; causative agent of typhus, spotted fever, and scrub typhus.

Riddell slide culture method a culture technique for the identification of fungi in which the sides of a small section (~2 cm × 2 cm) of agar medium are cut from a plate and placed on microscope slide that is on a bent U-shaped glass rod in a petri plate, inoculated with the test organism, and covered with a sterile glass cover slip so that following incubation the coverslip can be removed, stained, and viewed without disruption of the delicate structures; variation of the Harris method.

RIFA radioimmunofocus assay.

rifampicin antibiotic that inhibits DNA-dependent RNA polymerases by blocking the RNA chain initiation step; used as a selective agent in growth medium.

RISA rRNA intergenic spacer analysis.

risk assessment the characterization of the potential adverse effects of exposure to environmental hazards; steps in the process include hazard identification, dose-response assessment, exposure assessment, and risk characterization.

risk-based prevention the targeting of individuals who are at increased likelihood for adverse health effects; in contrast to population-based prevention.

risk characterization the area of risk assessment that encompasses the description of the nature and magnitude of risk and uncertainty of environmental exposure.

risk factor a feature or characteristic of an individual that is associated with an increased probability of acquiring a particular disease.

Risk management program (RMP) a program of the United States Environmental Protection Agency to provide a system for submission of risk management plans for the prevention of chemical accidents, injury and illness.

RMP risk management program.

RNA ribonucleic acid.

RNA polymerase an enzyme that uses an antiparallel DNA strand as a template to synthesize RNA in the 5′ to 3′ direction.

RNase ribonuclease.

Robbins device a device used to study biofilms *in situ*; composed of a pipe or tube with removable bolts in the walls that can be assayed for microbial growth and composition.

RODAC plate replicate organism detection and counting culture plate.

root nodule a growth on a plant root that contains nitrogen-fixing bacteria; formation of the nodule is recognized as involving 6 steps: recognition and attachment of the bacterium to root hairs, excretion of nod factors, invasion of the root hair and formation of an infection thread, travel to the main root via the infection thread, formation of bacteroids within the plant cells, and formation of the mature nodule.

Rose Bengal agar a selective culture medium used for the growth of yeast and molds; the presence of the rose bengal dye inhibits the growth of bacteria and restricts the size of the mold colonies; medium must be protected from sunlight as it will degrade; may be amended with chloramphenicol (Rose Bengal Chloramphenicol Agar).

rot plant pathology term used to describe the symptom of some plant diseases that result in death of cells or tissue.

rotary drilling a drilling technique used to obtain samples in deep-sediment and rock subsurface environments that requires the use of drilling fluids; the collected material is altered by the introduction of the drilling fluid resulting in changes in chemical and microbiological composition.

rotating annular bioreactor (RAB) a continuous culture device used for the study of biofilms under near steady-state conditions; concentric cylinders are arranged so that one is fixed and the other can rotate providing measurements of shear and friction that change with developing microbial populations.

rotavirus a genus of double-stranded RNA viruses in the family Reoviridae; believed to be the most significant cause of severe gastroenteritis in young children worldwide; significant cause of childhood mortality in developing countries; shed in very high numbers in feces of infected individuals; transmitted by the fecal-oral route.

rotor housing that holds individual sample containers in a centrifuge.

route of entry/route of exposure path by which a foreign substance enters the body.

RpoS an alternative sigma factor, present in gram-negative bacteria, that regulates expression of several genes during environmental stress and entry into the stationary phase of growth.

rRNA ribosomal RNA.

RTE ready-to-eat foods.

16S rRNA ribosomal RNA that is found in the small (30S) subunit of prokaryotic ribosomes; involved in the conversion of genomic DNA sequences into functional proteins, and because the16S rRNA molecule must maintain certain structural characteristics to preserve its function, the DNA sequence that encodes the 16S rRNA molecule is highly conserved; by sequencing the divergent regions in the 16S rRNA genes and comparing these sequences to a database of known sequences, identification of a bacterial isolate can be achieved.

23S rRNA one of two chains of ribosomal RNA that comprise the large (50S) subunit of prokaryotic ribosomes.

rRNA intergenic spacer analysis (RISA) a method in which the region between rRNA genes of different organisms is studied to assess the structure of bacterial communities in soils.

r strategists microorganisms that rely on high productivity rates for continued survival within a community and prevail in environments that are not resource limited.

RT-PCR reverse transcriptase polymerase chain reaction.

rugose coarsely wrinkled in appearance.

Ruminococcus bacterial genus; cellulolytic species are important in the degradation of plant cell wall polysaccharides in the rumen and hindgut of mammals.

rusts members of the Basidiomycetes with a complex life cycle of five spore types requiring two different host plants; ubiquitous organisms representing more than 100 genera and 5000 species that do not grow on laboratory media but are readily observed on spore trap air samples; isolated from grasses, flowers, and trees; associated with Type I allergies.

Rutgers process an aerated pile composting method for organic wastes in which air is injected into the pile of material for oxygenation and temperature control; in contrast to the Beltsville method.

S

σ lowercase Greek symbol for sigma that is used to denote standard deviation.

σ^2 lowercase Greek symbol for sigma squared that is used to denote variance.

σ^B sigma B.

Σ uppercase Greek symbol for sigma used to denote the summation of a series of numbers.

Sabouraud dextrose agar (SDA) a fungal culture medium containing 40 grams of dextrose per liter; generally used for the isolation of medically important organisms.

Saccharomyces fungal genus; round to oval, multilateral budding yeast cells with short pseudohyphae.

Saccharomyces carlsbergensis fungal species; a brewery bottom yeast that is used in production of lager beer.

Saccharomyces cerevisiae fungal species; a brewery top-fermenting yeast that is used in the production of ale.

SAED selected-area electron diffraction.

Safe Drinking Water Act (SDWA) established in 1974 provides for the development of drinking water standards by the US Environmental Protection Agency.

safety glasses/safety goggles eye protection for laboratory workers that generally consist of clear plastic or other material extending beyond that provided by regular eyewear.

safety officer individual assigned to document and maintain safety protocols in the laboratory.

safety training detailed instruction of workers on the correct handling of materials in the laboratory.

safranin dye used as the counterstain in the Gram stain procedure to impart a pink to red coloration to bacteria.

Salmonella bacterial genus; usually motile, gram-negative bacilli that produce acid and gas from glucose with a characteristic production of hydrogen sulfide; common inhabitants of the intestinal tract of humans and lower animals; pathogenic species are the causative agent of typhoid fever, gastroenteritis, and sepsis.

Salmonella enterica bacterial species; food-borne pathogen transmitted via the ingestion route of exposure through the consumption of raw tomatoes.

Salmonella paratyphi bacterial species; waterborne and food-borne pathogen transmitted via the ingestion route of exposure.

Salmonella typhi bacterial species; waterborne pathogen transmitted via the ingestion route of exposure; causative agent of typhoid fever.

Salmonella typhimurium bacterial species; food-borne pathogen transmitted via the ingestion route of exposure; most common causative agent of salmonellosis in the United States; used in the Ames test to determine toxicity of chemicals.

***Salmonella typhimurium* WG49** a serotype of *Salmonella choleraesuis* that is used as a host for male-specific bacteriophages.

salmonellosis self-limiting gastrointestinal disease caused by the ingestion of *Salmonella*; onset of symptoms occurs after several days due to the amplification of the bacteria in the intestine.

sample the material to be analyzed or the action of collecting material to be analyzed.

sample collection gathering a portion of material that is representative of the whole without positively or negatively altering the populations present and without contaminating the material with foreign substances.

sample incorporation the inoculation of a small volume of sample into a large volume of concentrated test medium.

sampling stress physical conditions that occur during collection of a sample that result in loss of viability of the microorganisms in the sample.

sanitary landfill facility for the disposal of refuse in which the material is covered daily with a layer of soil using one of three methods, area method, the trench method or the ramp variation.

saprophyte/saprophytic an organism that utilizes organic matter from dead organisms for nutrient.

SARA Superfund Amendments and Reauthorization Act.

sarciniform packet-like.

SAS Super 90 sampler a handheld, battery-powered impactor sampler designed with a single stage containing multiple round inlets for the collection of bioaerosols at a flow rate of 85 liters/min. onto an agar-filled dish for follow-on culture analysis.

satratoxin G macrocyclic trichothecene mycotoxin; associated with *Stachybotrys*.

satratoxin H macrocyclic trichothecene mycotoxin; most common toxin associated with *Stachybotrys*; can depress immunological responses; not classified by IARC for human or animal carcinogenicity; epidemiological data suggest a higher rate of upper respiratory tract and lung cancers in granary workers and food handlers with inhalation exposure to high concentrations of fungal products.

SBYR Gold a nucleic acid-staining fluorescent dye used with epifluorescent microscopy for detecting single-stranded and double-stranded DNA or RNA.

scab localized lesion that is generally slightly raised or sunken; symptom of some plant diseases caused by microorganisms.

scabrous rough in appearance with short projections.

scale-up moving from a laboratory or demonstration study to a full-scale, operational process.

scanning electron microscope/scanning electron microscopy (SEM) microscopic method using the scattering of electrons on a sample treated with a thin film of gold or other heavy metal to visualize surface structure.

Schloesing, J.J. reported microbial nitrifying activity in 1877–1879 with Muntz; demonstrated that the ammonium in sewage was oxidized to nitrate when passed through a sand column, an activity that was eliminated when chloroform was introduced and then restored when a soil suspension was used as an inoculum.

Schneeberg disease lung cancer of uranium miners; associated with exposure to high concentrations of radon and *Aspergillus flavus* in underground mines.

Schwann, Theodor 18th century scientist whose experiments implicated airborne microorganisms in the spoilage of heat-sterilized substances; concluded independently of F. Kuntzing, but in the same year (1837), that alcoholic fermentations were caused by yeasts.

scientific investigation plan experimental design.

sclerotia plural of sclerotium.

Sclerotina sclerotiorum filamentous actinomycete; phytopathogen that acidifies its ambient environment by producing and secreting oxalic acid.

sclerotium a hardened mass of fungal hyphae.

scope of work (SOW) details of the goals and objectives, experimental design, and protocols used during the conduct of a project.

Scopulariopsis fungal genus; colony on malt extract agar is powdery to funiculose in texture with a central tuft; white becoming brownish-rose with a pale to brown reverse; rose-brown rough conidia are borne on conidiogenous cells that arise singly on aerial hyphae or on branched conidiophores in groups of 2–3; isolated from soil, wood, straw, dead insects, house dust, carpets, wallpaper, paper, and painted surfaces; may produce harmful VOCs when growing on arsenic compounds in wallpaper and tapestry.

scorpioid coiled like the tail of a scorpion.

scotochromogenesis formation of pigment only when the microorganism is cultured in dark; used in the classification of some *Mycobacterium* spp.; in contrast to photochromogenesis.

scotophobotactic response a phototaxis-type response in which bacterial cells in a wet mount preparation reenter the lighted area of a microscopic field from the dark periphery perhaps due to changes in energy generation.

screening the analysis of samples using less expensive, less precise, and/or less quantitative methods to reduce the sample size before a more rigorous methodology is used.

scutellate shaped like a small shield.

SDA Sabouraud dextrose agar.

SDS sodium dodecyl sulfate.

SDS–PAGE sodium dodecyl sulfate–polyacrylamide gel electrophoresis.

SDWA Safe Drinking Water Act.

SE standard error of the mean.

secondary ion-mass spectrometry (SIMS) an analytical technique used for the characterization of surfaces that employs an ion beam to sputter atoms and ions from the surface that are then trapped and analyzed by mass spectrometry.

secondary metabolite material produced at the end of the growth phase and during the stationary phase of the microbial life cycle as a by-product of metabolism.

secondary prevention the identification of people who have developed a disease while it is still at an early stage by screening and early intervention; in contrast to primary prevention.

secondary productivity the heterotrophic utilization of substrate; in contrast to primary production.

secondary spread refers to spread of infection to individuals who were not exposed to the originally contaminated source, but rather to individuals who were infected by the original source.

secondary standard a regulatory pollution limit value for a criteria air pollutant or drinking water contaminant that is based on environmental effects; in contrast to a primary standard.

secondary succession establishment of a new population within a habitat that has previously been colonized by the population of the primary succession.

secondary wastewater treatment aerobic and anaerobic processes mediated by microorganisms that are used to decrease the level of organic material following primary wastewater treatment; anaerobic processes are conducted in sludge digesters or bioreactors while aerobic processes utilize trickling filters or activated sludge.

second-order epidemic synoptic spread of a plant pathogen occurring for several years and over several thousand kilometers in area; in contrast to zero-order epidemic and first-order epidemic.

sedimentation the settling of particulate matter in a column of liquid.

selection the growth of a particular microbial population due to the presence or absence of a nutrient or physical condition.

selection bias the introduction of a systematic error due to the manner in which the test and control populations are selected; in contrast to surveillance bias and misclassification bias.

selective agent/selective culture medium a reagent-amended preparation that permits the growth of specific microorganisms and suppresses the growth of others.

self-limiting descriptive of a disease that does not require antibiotic treatment.

SE_M standard error of the mean.

SEM scanning electron microscopy.

seminested PCR a modification to the polymerase chain reaction amplification method, in which the initial PCR is followed by a second PCR in which a second set of primers, one of which is identical to that used in the initial PCR and the other of which is designed to anneal to a sequence internal to the product of the initial PCR.

sensitivity the detection capability to correctly identify the presence of an analyte of interest.

Sephadex a commercial product of crosslinked dextran gel beads used in the separation of materials.

Sepharose a commercial product of agarose gel prepared without charged polysaccharides.

septate a condition in which a demarcation exists within a structure that segments the unit into divisions.

septic tank simple anaerobic wastewater treatment method in which solids settle to the bottom of a tank and clarified effluent is distributed over a leach field; generally used to treat waste from a single household or business or a small group of residences.

septa plural of septum.

septum a crosswall within a structure.

SERC State Emergency Response Commission.

serial dilution method to sequentially reduce the concentration of microorganisms in a liquid suspension; a 10-fold series is accomplished by pipetting 1 ml of a microbial suspension into 9 ml of diluent.

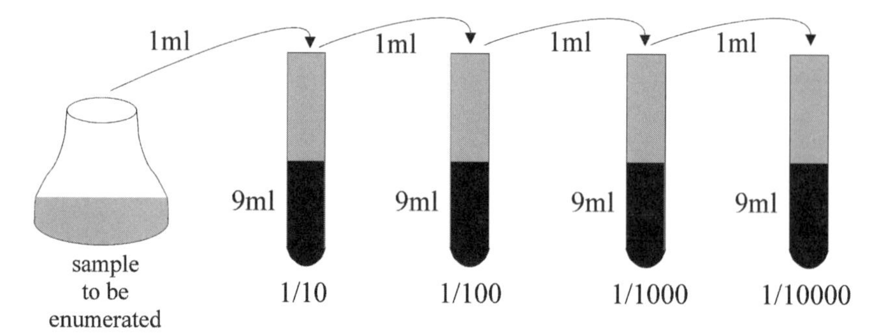

Serial dilution

serogroup/serotype a subdivision of a species or subspecies of organisms, determined on the basis of antigenic characteristics; classification of organisms based on antigenic characteristics.

serological pipet calibrated pipet for the transfer of liquids over a range of measurements; in contrast to a volumetric pipet.

serotype an antigenic variation of a microbial species.

serous fluid thin, watery serum-like liquid.

Serratia bacterial genus; member of the Enterobacteriaceae; small, gram-negative coccobacilli found in soil and water, and occasionally in humans.

Serratia marcescens bacterial species; most strains produce colonies in culture with an orange-red pigment, prodigiosin, but some are nonpigmented; has been used as a tracer to demonstrate atmospheric dispersal of bioaerosols.

sessile refers to a structure that is attached to a surface by a base rather than by a stalk.

seta/setae bristle/bristles.

settled dust sample/settled dust sampling the collection of particulate from surfaces for identification and enumeration of microorganisms; generally collection is accomplished using a vacuum collection device such as a cassette attached to a vacuum pump or a specialized adaptor on a vacuum cleaner; samples are analyzed by direct microscopy or the sample is processed for culture.

sewage fungus term used in wastewater treatment to describe the slime material that accumulates on rocks or other solid materials resulting from the growth of filamentous bacteria such as *Sphaerotilus natans*; these organisms participate in wastewater treatment by oxidizing organic material.

sex pilus a hairlike appendage possessed by many bacteria that serves as the attachment site for male-specific bacteriophage, also joins bacteria together to allow the transfer of DNA from one cell to the other; F pilus.

SFG rickettsiae spotted fever group *Rickettsia*.

Shannon-Weaver Index of Diversity ($\bar{H}$) widely used formula to express both species richness and relative species abundance.

$$\bar{H} = C/N\,(N \log_{10} - \Sigma n_i \log_{10} n_i)$$

$C = 2.3$
N = # of individuals
n = # of individuals in the ithspecies

sheathed bacteria term for bacteria whose cells occur within a filament that enables the organisms to attach to solid surfaces and serve to protect the cells from predators and parasites.

shelf life the time period in which a product is considered to exhibit expected properties.

Shelford's Law of Tolerance theory that each organism in an ecosystem requires a complex set of conditions for survival and growth and these conditions must remain within the tolerance limits for an organism to thrive or it will be eliminated.

Shewanella putrefaciens **MR-1** bacterial species; a gram-negative, metal-reducing microorganism that uses a wide variety of terminal electron acceptors for anaerobic respiration.

Shiga-like toxin a group of toxins that are similar in structure to Shiga toxin.

Shiga toxin a bacterial toxin produced by *Shigella dysenteriae* that blocks protein synthesis in eukaryotic organisms.

Shigella bacterial genus, member of the family Enterobacteriaceae; gram-negative, chemoorganotrophic, straight, nonmotile; anaerogenic bacteria that ferment dextrose but not lactose; pathogenic to humans.

Shigella dysenteriae bacterial species; human pathogen with potential environmental exposure via aerosols generated during wastewater treatment practices as well as contaminated drinking water, recreational water, and food.

short-term emergency limits (SPELs) former term for short-term public emergency guidance levels.

short-term public emergency guidance levels (SPEGLs) concentrations that are suitable for unpredicted, single, short-term emergency exposure of the general public; in contrast to emergency exposure guideline levels.

sick building syndrome the excessive reporting of symptoms such as headache, lethargy, cognitive or upper respiratory problems, and flu-like complaints associated with occupancy of a specific indoor environment, but a causative agent has not been identified; in contrast to building related illness.

Siderocapsa bacterial genus; member of the Siderocapsaceae; spherical cells that are embedded in a capsule encrusted with iron or manganese oxides; found in fresh water.

Siderocapsaceae bacterial family; members characterized by their ability to oxidize iron or manganese and deposit the metal oxides in capsules or in extracellular material.

Siderococcus bacterial species; member of the Siderocapsaceae; spherical cells that deposit iron but not manganese and are not coated with iron oxides; distributed in freshwater and sediments.

siderophore a natural iron chelator that binds iron present at low concentrations.

sigma B (σ^B) an alternative sigma factor, present in gram-positive bacteria, that regulates expression of several genes during environmental stress (e.g., heat, acid, salt, and ethanol) and entry into the stationary phase of growth.

signature a chemical or biochemical profile that is unique to a particular microorganism.

sign an objective indication of a condition that is measurable, such as the temperature of a patient; in contrast to symptom.

silica gel inorganic solidifying agent; often used when culturing ammonia oxidizing nitrifying bacteria; also used as a desiccant.

siliquiform spindle-shaped.

simple unbranched.

simple regression statistical analysis to illustrate the relationship between two variables in which both the independent and dependent variables are interval data; in contrast to multiple regression.

simple sequence repeat (SSR) sections of DNA composed of 10 to 60 tandem repeats of two nucleotides (e.g., AGAGAGAGAGAGAGAGAGAG); analyses of these sections can be used to differentiate between individual organisms.

SIMS secondary ion mass spectrometry.

simulation model mathematical relationships use to reflect and predict detailed changes within a system; in contrast to a theoretical model.

single agar layer method a method used to detect and enumerate bacteriophages in which the sample, host, and molten agar are combined, then poured into a petri plate, cooled, and then incubated to allow infection of the host and the appearance of plaques.

single-cell protein a food or medical product that is derived from microbial activity.

single-strand-conformation polymorphism (SSCP) a mutation, as small as a single base shift, in a DNA sequence that can be detected by using polymerase chain reaction amplification to amplify the sequence, visualizing the product on a high quality gel and comparing it to the known nonmutated sequence.

Sinorhizobium bacterial genus; gram-negative, nitrogen-fixing, motile bacillus that lives in a symbiotic relationship with legumes (e.g., beans, clover, alfalfa, and peas) forming root nodules.

sinuous wavy or serpentine in appearance.

SIP scientific investigation plan.

SKC BioSampler an impingement sampler designed for the collection of biological particles into a liquid collection medium; utilizes three narrow inlets to direct the incoming air toward the perimeter of the collection vessel resulting in a swirling of the liquid in contrast to the unidirectional orifice of the AGI-30 sampler.

skew the symmetry of a curve.

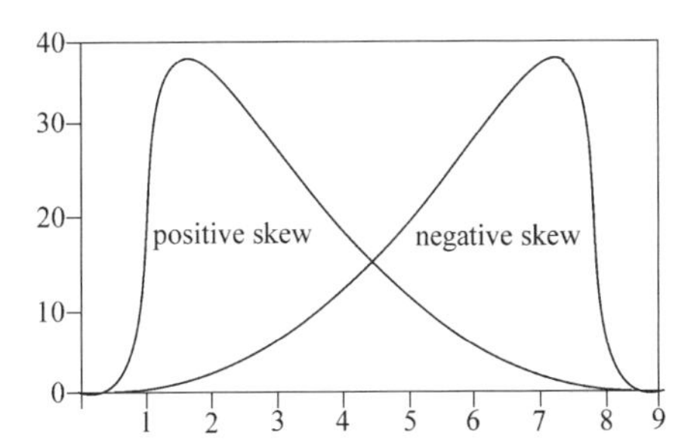

skew left negative skew.

skew right positive skew.

slant rotor centrifuge housing that holds the sample containers at a fixed angle.

S-layer an outer wall layer of protein or glycoprotein that is found in many prokaryotic organisms.

slide culture techniques methods used to culture fungi for identification with light microscopy; Riddell method and Harris method.

slime layer a glycocalyx arrangement around a cell that is easily deformed and does not exclude particles.

slime molds nonphototrophic eukaryotes that phenotypically resemble fungi and protozoa; primarily found on decaying plant material and soil where they rely on the ingestion of bacteria by phagocytosis for food.

slit inlet rectangular-shaped opening.

slit sampler impactor sampler that utilizes a rectangular shaped inlet nozzle.

slit-to-agar (STA) sampler an impactor sampler that collects bioaerosols through a slit onto a rotating 150 mm diameter agar surface; capable of providing data on the concentration of airborne bacteria and fungi with time discrimination as the agar surface rotates at a predetermined rate.

slot blot a variation of the dot blot procedure in which the solution to be analyzed is placed in slots on a matrix, rather than being spotted on the matrix.

sludge solid product of wastewater treatment.

sludge cake compacted solid product of domestic wastewater treatment.

SMAC agar sorbitol-MacConkey agar.

smoke mixture of particles and gases resulting from incomplete combustion.

smoke pencil narrow cylindrical device that emits a fine cloud of smoke; used to visualize the flow of air in a defined area.

smut members of the Basidiomycetes with a complex life cycle of two spore types requiring a living host plant when in the teliospore phase, but will grow on laboratory media during the yeast phase; ubiquitous organisms representing 50 genera and 950 species; readily observed on spore trap air samples; difficult to distinguish teliospores from myxomycetes; isolated from cereal crops, grasses, flowering plants, and other fungi; associated with Type I allergies.

Snow, John British physician who applied the principles of epidemiology to determine the waterborne transmission of *Vibrio cholerae* in London in 1855.

sodium dodecyl sulfate (SDS) an anionic detergent that denatures proteins; used in gel electrophoresis to permit independent migration of polypeptides according to size or apparent mass.

sodium thiosulfate ($Na_2S_2O_3$) a dechlorinating agent added to aqueous samples to neutralize residual chlorine and other halogens and minimize bactericidal action.

soft agar preparation of semisoft agar (approximately 0.7% agar w/v) used in plaque assay for bacteriophage.

soft-rot phytopathogen bacterium that infects plant stems and shoots but not leaves.

soil column the use of a column containing sieved soil or a soil core to assess the movement, concentration or composition of chemicals or microbial populations through the column.

soil extract agar culture medium prepared using garden soil that stimulates the sporulation of some fungi.

soil moisture the amount of moisture present in a sample of soil; determined by weighing a sample before and after drying.

solfatara a hot, generally acidic, sulfur-rich environment inhabited by thermophilic Archaea.

soluble capable of being dissolved; usually dependent on the composition of the dissolving solution and the temperature.

soluble ultrafilter filtration matrix comprised of gelatinous material that is placed in a growth medium or liquefied during analysis of the retained material.

solum the combined A horizon and B horizon.

solute that which is dissolved.

somatic phage a phage whose attachment site is the cell wall of a bacterium; in contrast to male-specific phage.

sonicator/sonication disruption of material using high frequency sound.

SOP standard operating procedure.

Soper, George epidemiologist who established the connection of "Typhoid Mary" as the carrier of *Salmonella typhi* in several outbreaks of typhoid in the early 20th century.

sorbitol-MacConkey (SMAC) agar used for the isolation of *Escherichia coli* (sorbitol-positive colonies) from stool samples of infected animals and humans.

Sordelli's method a drying technique for the preservation of microorganisms that uses an inner tube containing a microbial suspension in horse serum and an outer tube containing P_2O_5 that is sealed with a vacuum.

source the place or object responsible for the release of a pollutant or other substance.

Southern, E.M. developer of the Southern blot hybridization procedure.

Southern blot hybridization method in which DNA is in a gel matrix and the DNA or RNA is the probe in contrast to the Northern blot hybridization method.

SOW scope of work.

sp. abbreviation for species that is placed after a genus name; used to denote a single species; in contrast to spp.

SP a single-stranded RNA, male-specific bacteriophage in the Leviviridae family.

SPAGE sodium dodecyl sulfate–polyacrylamide gel electrophoresis.

Spallanzani, Lazzarro (1729–1799) Italian naturalist who attempted to dispell the notion of spontaneous generation in the early 18th century by demonstrating that heat destroyed microorganisms and sealing of containers prevented spoilage.

specialized transduction transfer of DNA from one bacterium to another in low frequency using viruses; occurs in temperate viruses with DNA from a specific region of the bacterial host chromosome integrated directly into the virus genome, usually replacing some of the virus genes and rendering the virus defective; because the genes are incorporated into the virus genome the DNA may be integrated into bacterial host chromosome during lysogenization or the DNA may be replicated in the recipient cell as part of a lytic infection; in contrast to generalized transduction.

species taxonomic classification of microorganisms below the genus level; a collection of strains that share the same major properties; as a Latin or Greek derivation species names are written in italics.

species diversity indices mathematical formulas used to express biological diversity by relating the number of species and the relative importance of individual species.

species richness (d) the variety component of species diversity expressed by simple ratios between the total number of species and total numbers of organisms, but not the number of organisms within each species; in contrast to equitability.

$$d = \frac{S-1}{\log N}$$

d = species richness
S = # of species
N = # of individuals

specific gravity ratio of the mass of a material to the mass of an equal volume of water, generally measured at 4°C.

specificity the ability to distinguish the analyte of interest from other materials.

spectrophotometer instrument with a prism or diffraction grating to generate incident light in a narrow band of wavelengths for the measurement of the absorbance or transmission of light by a liquid sample; used to determine the optical density of a liquid culture medium when performing a growth curve; used to quantify macromolecules when used in the ultraviolet light range.

SPEGLs short-term emergency guidance levels.

SPELs short-term emergency limits.

Sphaerotilus bacterial genus; nutritionally versatile sheathed bacteria commonly isolated from streams and brooks containing abundant leaf litter, and surfaces receiving effluent with a high organic content; sheath can be encrusted with iron oxides.

Sphaerotilus natans bacterial species; often termed the sewage fungus due to its presence in sewage effluent and polluted flowing waters.

spherical aberration defect in a microscopic image due to the passage of light through thinner portions of a convex-convex lens, resulting in focusing in slightly different planes than the light passing through the wider center portion of the lens; corrected by the use of an apochromatic lens; in contrast to chromatic aberration.

spherule a thick walled spherical structure characteristic of *Coccidioides immitis* that is formed within host tissue.

Sphingomonas bacterial genus; member of the family Pseudomonaceae; gram-negative, chemoorganoheterotrophic organisms widely distributed in soils, sediments, and water that are capable of degrading contaminants present in activated sludge.

Spingomonas chlorophenola bacterial species; important in the bioremediation of pentachlorophenol-contaminated sites.

Sphingomonas paucimobilis bacterial species; potential environmental exposure to humans via aerosols generated during wastewater treatment practices.

spinescent having spines.

spinulose having small spines.

spiral hyphae a corkscrew or flattened spiral-shaped formation at the terminal end of some fungal hyphae.

spirillum spiral-shaped.

Spirochaeta bacterial genus; member of the Spirochaetales that are free-living anaerobes.

Spirochaetales bacterial order; slender, motile unicellular bacteria that are helically coiled.

Spiroplasma bacterial genus; member of the order Mollicutes; some species are phytopathogens, causative agents of disease in citrus, grasses, and corn.

splash-dispersed fungal spores airborne fungal spores that are passively released from plant surfaces as the result of raindrop or irrigation splash; the formation of a polysaccharide mucilage layer from the plant prevents the dispersal and protects the spore from desiccation during dry weather but it dissolves during water events permitting the release of the spores; formation of large droplets prevents long-range dispersal resulting in dissemination downward from younger to older leaves of the plant.

split/split cell the division of the cells in cell culture from one flask to multiple flasks for the propagation of new cells; a 1:4 split means that the cells from a single flask would be split into 4 flasks.

spontaneous generation the theory that living organisms could originate from nonliving materials that was disproved by Louis Pasteur in 1864 with his classic experiment using a goose-necked flask.

sporadic case report of an isolated incidence of disease.

sporangia plural of sporangium.

sporangiferous descriptive of a structure that bears sporangia.

sporangiole a small sporangium with a few sporangiospores.

sporangiophore a segment of specialized fungal hyphae that produces a sporangium.

sporangiospore walled, asexual spore produced in a sporangium that is characteristic of the Zygomycetes.

sporangium an enclosure of endogenous fungal spores that were not derived from the containing structure.

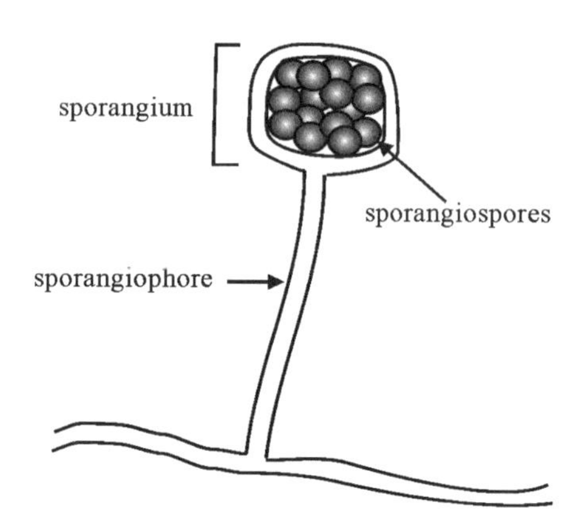

spore a reproductive structure of fungi and bacteria; fungal spores are single cells that may be produced in response to adverse conditions or as a result of asexual or sexual reproduction with dissemination via wind, and many have distinctive morphology that can be used to distinguish some genera; bacterial spores are more correctly termed endospores.

spores/m^3 measurement of the number of fungal spores enumerated by nonculture-based methods per cubic meter of air sampled.

spore strip an indicator strip containing endospores of a *Bacillus* species that serves as a quality control check for the operation of an autoclave.

spore trap air sampler an impactor sampler used to collect airborne particulate, including fungal spores, pollen, and debris; generally these samplers operate at a fixed flow rate designed by the manufacturer with the particulate being deposited onto a sticky surface for analysis using microscopy.

Sporobolomyces fungal genus; yeast that forms buds on short stalks; colony on malt extract agar is shades of peach, pink, and salmon in color and characteristic satellite colonies are present around the main colony due to the release of ballistospores; active spore discharge during periods of high humidity as the absorption of moisture is required to build up release pressure; airborne spores generally more abundant at night or predawn hours are identified as basidiospores on spore trap samples; ubiquitous; approximately 10 species; most successful fungus in the phyllosphere where it is isolated from tree leaves, soil, rotting fruit, and it can be found on a variety of wet indoor substrates; implicated as a cause of dermatitis and associated with Type I allergies and Type III hypersensitivity.

sporocarp fruiting body; spore-bearing organ characteristic of some fungi.

sporodochia plural of sporodochium.

sporodochium a compact mass of interwoven conidiophores.

Sporothrix schenckii fungal species; a diphasic fungus with a hyaline to gray or dull brown floccose to velvety or lanose colony on malt extract agar with a gray or black reverse and submerged hyphae and ovoid, pyriform, to fusiform conidia (2–6 μm) and black perithecia at 30°C; soft, white to cream colored irregular yeast colonies are cultured at 37°C; causative agent of chronic subcutaneous lymphatic mycosis in humans, also called sporotrichosis resulting from inoculation in broken skin while handling soil; worldwide distribution independent of soil type or climate.

sporotrichosis a subcutaneous infection caused by *Sporothrix schenckii*; characterized by swollen lymph nodes near the point of inoculation and is generally seen in gardeners who become scratched and receive wounds exposed to soil.

sporozoa nonmotile protozoa.

sporulating/sporulation the production of spores, generally denoting the formation of fungal conidia.

spp. abbreviation for species that is placed after a genus name; used to denote more than one species; in contrast to sp.

spread plate/spread plating a method in which a microbial suspension of a known volume, generally 100 μl, is evenly distributed onto a solidified agar surface (e.g., agar-filled petri plate) for culture analysis.

sputum lysis buffer a suspension solution used in the immunomagnetic separation detection of Helicobacter.

squamulose scaly in appearance.

SRB sulfate reducing bacteria.

SSCP single-strand-conformation polymorphism.

ssDNA single-stranded DNA.

SSR simple sequence repeat.

ssRNA single-stranded RNA.

STA slit-to-agar sampler.

Stachybotrys fungal genus; colony on malt extract agar is powdery in texture, initially white to gray turning black to blackish green with a colorless reverse; conidiophores simple or branched becoming brown with age; phialides ellipsoidal becoming black; aggregates of dark spores produced in a slimy mass compared to the spore chains observed for Memnoniella; ubiquitous; approximately 15 recognized species.

Stachybotrys atra former terminology for *Stachybotrys chartarum*.

Stachybotrys chartarum fungal species; white colony on malt extract agar turning dark with masses of slimy ellipsoidal, dark olivaceous conidia on simple or irregularly branched conidiophores; potential for production of mycotoxins; isolated outdoors from the soil, decaying plant debris such as leaf litter and cellulose-containing material (e.g., hay, straw), indoors isolated from cellulose-containing building materials (e.g., wallboard, wicker, straw baskets, paper products and jute); Aw 0.94; colonies not readily isolated on routine mycological medium in the presence of *Penicillium* and other commonly isolated fungi; use of cellulose-based culture medium may enhance likelihood of isolation.

stalk an elongated structure that serves to attach a microbial cell to a surface.

standard deviation (S, σ) the relationship of individual values to the mean; the square root of the variance or average deviation of each number; the mean of all the numbers in a set of data expressed in the same units as the original measurement.

$$s = \sqrt{\frac{\Sigma(X-\overline{X})^2}{n-1}} = \sqrt{\frac{\Sigma x^2}{n-1}}$$

standard error of the mean (SE, SE$_M$) relationship of the sample size and the standard deviation that is used to show how close the means from repeated samples are to the population mean.

Standard Methods for the Examination of Water and Wastewater a compilation of approved methods for the analysis of water and wastewater published by the American Public Health Association, the American Water Works Association, and the Water Pollution Control Federation; first published in 1905, the current edition details the examination of physical and aggregate properties, metals, inorganic nonmetallic constituents, aggregate organic constituents, individual organic compounds, radioactivity, toxicity, microorganisms, and other biological substances.

standard operating procedure (SOP) a written method, usually compiled in a notebook, that is followed for the conduct of a specific task.

standard plate count (SPC) heterotrophic plate count.

standard score (z, Z) the distance of a value from the mean expressed in standard deviation units; also termed zee score, z-score, and zed score.

Standards for the Use or Disposal of Sewage Sludge, 1993 regulation established by the United States Environmental Protection Agency to set minimum quality standards for the land application of sewage sludge; also known as the Part 503 Rule.

standard temperature and pressure (STP) 25°C (298 K) and 1 atmosphere (760 mmHg).

Stanier, Roger student of van Niel who demonstrated the versatility of microorganisms in the degradation of complex organic compounds.

staphylococcal food poisoning gastroenteritis resulting from the ingestion of foods containing enterotoxin produced by *Staphylococcus aureus*; short incubation period of 2–6 hours with disease lasting approximately 24 hours and recovery in 1–2 days; associated with cream-filled pastries, puddings, and other food processed with a minimum of heat or improper refrigeration.

Staphylococcus bacterial genus; facultatively anaerobic, gram-positive, nonmotile chemoorganotrophic cocci that occur singly, in pairs, or clusters; natural populations are associated with the skin and mucous membranes of humans and animals and therefore can readily contaminate animal products such as meat, cheese and milk; also isolated from environmental sources such as soil, sand, dust, air and natural waters.

Staphylothermus bacterial genus; member of the Archaea; anaerobic hyperthermophiles that form aggregates of 1 μm spherical cells and have a temperature range of 65–98°C with an optimal growth temperature of 92°C; S^o is required for growth but oxidation of organic compounds is not coupled with S^o reduction; commonly distributed near hydrothermal vents.

static not in motion; inhibitory.

static pile composting a relatively slow, nonuniform method in which material to be decomposed is stacked in long windrows with scheduled turning for aeration and temperature adjustment; in contrast to aerated pile composting and continuous feed reactor methods.

stationary phase period of time in the growth curve of a microbial population in which there is no net increase or decrease in the number of organisms.

statistical significance the likelihood that an observed difference is a true difference.

STEC Shiga toxin-producing *Escherichia coli*.

stellate star-shaped.

STELs short-term exposure limits; a time-weighted average exposure for ≤15 min. period.

stem-leaf plot developed in 1977 as a way to transform interval data into a graph by finding the range, choosing an appropriate interval width that would result in 10–20 intervals, and then making a table that resembles a horizontal histogram with the "stem" being the most significant digit of the numbers and the "leaf" being the least significant digit.

Stemphylium fungal genus; large colonies with hyaline to brown hyphae; erect pigmented conidiophores that are darker and swollen at the apical end producing a single, pale to deep olive-brown, ovoid or ellipsoidal, muriform conidium with a longitudinal septum and several cross septa; associated with Type I allergies and evidence that the major allergen is Alt a I.

stepwise regression statistical analysis using multiple regression in which the variables are entered one at a time to determine how much is being gained with each variable; in contrast to hierarchical stepwise regression.

stereomicroscope instrumentation used to magnify an object to aid in identification of external structures; commonly used in mycology for the identification of fungal colonies.

stereoisomers two molecules that have the same structure but are arranged as mirror images.

sterigma a projection from a basidium that produces a basidiospore; formerly term for phialide and metula.

sterigmata plural of sterigma.

sterigmatocystin a carcinogenic mycotoxin that can be produced by *Aspergillus versicolor* and other species of *Aspergillus* when grown on certain substrates; precursor to aflatoxin.

sterile free of living organisms and viruses.

sterile mycelia vegetative structures without production of spores frequently due to inadequate culture conditions including inappropriate nutrients, temperature, light and dark cycling or ultraviolet light; often observed but not limited to the isolation of basidiomycetes on laboratory media.

sterilize/sterilization physical or chemical treatment that renders a material free of living organisms and viruses.

stilboid having a stalked head.

stinkhorns common term for members of the order Phallales; characterized by a gelatinous ooze and foul smell produced when the basidiocarp undergoes autodigestion, releasing basidiospores.

stipe stalk or stem.

stipple dot.

stir bar small magnetic rod used to assist in the mixing of liquid suspensions in a beaker or flask positioned on a stir plate.

stir plate electrically powered device designed to continuously mix a liquid suspension in combination with a stir bar that rotates 360° at a variety of speeds selected by the operator.

stochastic an event that is probabilistic in nature; contrast with deterministic.

stolon a segment of aerial hyphae or horizontal hyphae that produces rhizoides and sporangiophores; observed in some genera such as *Rhizopus*.

stomacher device for blending a sample with a liquid using paddles that are positioned to pulsate against a bag containing the material.

STP standard temperature and pressure.

strain taxonomic classification of microorganisms below species; a population of cells that are all descendants of a single cell.

stratified randomized trial a layering of study and control populations by variables that are considered important and then randomizing within each group.

stratosphere layer of air in the atmosphere that exists above the troposphere and below the ionosphere that contains a high concentration of ozone that absorbs utraviolet (UV) radiation, thereby protecting the earth from UV radiation, and acts as a barrier in the transport of airborne microorganisms to or from the troposphere.

Streptococcus bacterial genus; gram-positive, chemoorganotrophic cocci with fermentative metabolism that appear in pairs and chains with light microscopic analysis; species are pathogens and normal flora in humans and animals.

Streptococcus bovis bacterial species; fast-growing, amylolytic, acid-tolerant, gram-positive coccus that ferments starch primarily to lactic acid.

Streptococcus cremoris bacterial species; isolated from milk and milk products; used in the souring of milk and the production of some cheeses.

Streptococcus faecalis *Enterococcus faecalis*.

Streptococcus lactis bacterial species; isolated from milk and milk products; used in the souring of milk and the production of some cheeses.

Streptococcus mutans bacterial species; primary habitat is the tooth surface of humans, also isolated from the feces; shown to be cause disease in teeth of laboratory animals.

Streptococcus suis bacterial species; pathogen of pigs; associated with severe illness in humans.

Streptococcus thermophilus bacterial species; tolerates temperatures of 45°C.

Streptomyces bacterial genus; filamentous, gram-positive; member of the aerobic actinomycetes.

Streptomyces clavuligerus bacterial species; important in the production of antibiotics such as penicillin and the cephalosporins.

Streptomyces ipomoeae bacterial species; phytopathogen, causative agent of pox of sweet potato.

Streptomyces scabies bacterial species; phytopathogen, causative agent of potato scab disease.

streptomycin a broad spectrum antibiotic produced by species of *Streptomyces*.

striate having furrows or lines.

stringency the reaction conditions (e.g., temperature, pH, salt species and concentration) in molecular biology that determine the degree to which only perfectly complementary strands of nucleic acid anneal; at high stringency conditions, perfect complementarity is required for two strands to anneal; at lower stringencies, strands with mismatched bases may anneal.

Student's t-test statistical analysis to determine if two sets of data are different by comparing the means and used when the data in the two groups are not linked to each other; based on the ratio of the difference between groups to the standard error of the difference.

subacute toxicity testing a study, lasting between 5–14 days, used to determine the toxicity of a substance.

subaerial zone supralittoral zone.

subcentric slightly asymmetrical in growth.

subchronic toxicity testing a study, lasting between 15 days and 6 months, used to determine the toxicity of a substance.

subculture to make a new culture using materials (e.g., cells, microorganisms) from an existing culture; the transfer of a small amount of a microbial culture from an agar plate or broth to another agar plate or broth, usually using a pipette or an inoculating needle or inoculating loop.

subcutaneous beneath the skin.

subglobose almost spherical but with a slightly longer length than width.

subicle having tufted or matted mycelium.

subjacent immediately below.

subjective analysis observations that involve the opinion of the analyst such as color or odor; in contrast to objective analysis.

sublanose slightly wooly in appearance.

subpulverulent slightly powdery or dusty in appearance.

substrate material that serves as a nutrient for growth; material acted upon by an enzyme.

substrate utilization assay an analysis method that relies on the conversion of a specific compound in order to be recorded as a positive result.

subtilase subtilisin-like serine proteases; distributed in bacteria, Archaea, and eukaryotes.

subtilisin commercially valuable enzymes synthesized in some bacteria (e.g., *Bacillus subtilis*, *Bacillus licheniformis*, *Bacillus amyloliquefaciens*, and *Thermococcus kodakaensis*).

subtype/subtyping a genetic variant of an organism.

succession temporal sequence of populations within a habitat.

Sudan black B a dye used with light microscopy for the detection of fat droplets which appear dark while bacterial cytoplasm appears colorless.

sulfate-reducing bacteria a morphologically diverse group of bacteria that utilize sulfate as a terminal electron acceptor under anoxic conditions; currently 18 genera are recognized.

Sulfolobus bacterial genus; a hyperthermophilic member of the Archaea; when living in an anaerobic environment, members of this genus utilize elemental sulfur (S°) and a variety of organic compounds as electron donors with O_2 as the electron acceptor; when living in an aerobic environment, members of this genus grow as a chemolithotroph using H_2 as the energy source.

sulfur bacteria colorless bacteria that use sulfur compounds such as hydrogen sulfide (H_2S), elemental sulfur (S°), and thiosulfate ($S_2O_3^{-2}$) as electron donors; the oxidation of H_2S generally occurs in stages with sulfate (SO_4^{-2}) the final product, although elemental sulfur is often deposited within the bacterial cells and stored as an energy reserve until the H_2S is depleted.

sulfur cycle a biogeochemical process involving the transformation of sulfur through a series of oxidation states.

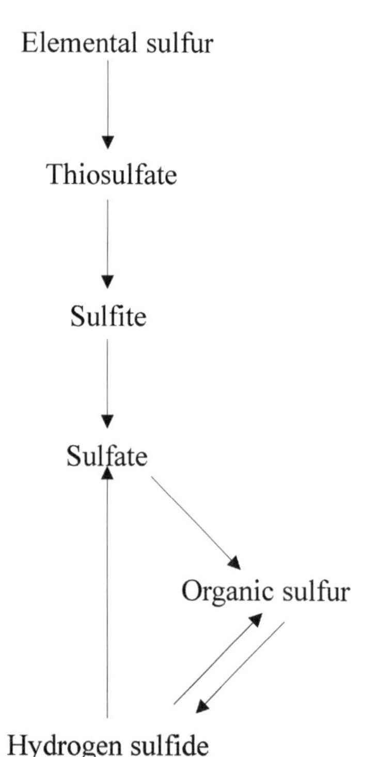

sulfur-reducing bacteria obligate anaerobes that utilize inorganic sulfur as terminal electron acceptors.

supralittoral zone interface region between the marine and terrestrial habitat that is washed with spray from the ocean and primarily populated by cyanobacteria and algae.

surface water treatment rule (SWTR) regulation promulgated by the United States Environmental Protection Agency in 1989 establishing that all public water utilities utilizing surface water or groundwater under the direct influence of surface water to treat the water to achieve 99.9% removal of *Giardia* and a 99.99% removal of viruses.

surrogate a similar material that is used in place of the real material; a microorganism that is used as a model for the behavior or presence of another microorganism or group of microorganisms (generally pathogens); the surrogate is used because it is easier, less expensive, and/or more rapid to detect than the organism of interest.

surveillance bias the introduction of a systematic error due to the increased attention that study populations are given compared to the general population; in contrast to selection bias and misclassification bias.

suspect carcinogen/suspected carcinogen classified by the United States Environmental Protection Agency as substances that may reasonably be expected to be carcinogens, there is limited evidence of carcinogenicity in humans or sufficient evidence of carcinogenicity in animals.

swab sampling method to assess microbial contamination on smooth or textured surfaces using a sterile cotton swab with analysis by culture or polymerase chain reaction amplification.

swing rotor hinged housing that rotates the angle of the sample containers during operation of the centrifuge.

SWTR surface water treatment rule.

SYBR Green a nucleic acid-staining fluorescent dye used with epifluorescent microscopy that has been shown to provide a similar fluorescence to that of YoPro-1 and a brighter fluorescence than 4′,6′-diamidino-2-phenylindole (DAPI) staining, but the fluorescence fades quickly.

sylvatic affecting rodents.

symbionts organisms that live in a mutualistic relationship.

symbiosis/symbiotic mutualism.

symbiosome a structure formed when plant cell membranes surround the bacteroid cell of nitrogen-fixing bacteria.

symptom a subjective observation of a condition, such as pain; in contrast to a sign.

synecology study of ecological systems with focus on the interactions between populations within the ecosystem; in contrast to autecology.

synergism/synergistic relationship of two microbial populations in which both populations benefit from the relationship although each population is capable of surviving in the environment alone; permits activities that neither population could perform alone such as synthesis of a product or degradation of a substrate; also called protocooperation.

synoptic spread dispersal of a crop pathogen over several years and over an area several thousand kilometers in size including coverage of a whole continent; in contrast to microscale and mesoscale spread.

syntopic occupying the same habitat.

syntrophy relationship of two or more microbial populations in which together they can degrade a substance in a series of reactions that cannot be accomplished by one of the populations alone.

systemic occurs at locations other than the site of exposure.

T

Taenioella fungal genus; produces dark brown conidia that may be collected in spore trap air samples, transparent tape samples, and from bulk material in indoor environments, but is not cultured on laboratory media; present on leaves and wood in nature, on plants and lumber in the indoor environment.

take-all disease a root disease of wheat and barley caused by *Gaeumannomyces graminis var. tritici*.

tangential flow filtration (TFF) a high efficiency ultrafiltration method used to collect viruses in water samples onto a filter matrix; requires preremoval of plankton and other suspended material to minimize the concentration of inhibiting substances.

tape lift technique to gently collect aerial hyphae and conidia (spores) from a fungal colony using transparent tape with analysis by light microscopy; terminology used for assay of fungal colonies in culture in contrast to transparent tape sampling used for surface sampling of building materials and furnishings in indoor environments.

tape sampling transparent tape sampling.

target organ the tissue where an adverse health effect occurs following exposure.

target organism the organism that is the subject of investigation.

Taq *Taq* polymerase.

TaqMan PCR assay a polymerase chain reaction amplification methodology that utilizes the 5′ to 3′ nuclease activity of *Taq* polymerase to digest a probe that binds to a region of DNA internal to the two primer binding sites. The probe consists of a fluorescent reporter dye at the 5′ end and a fluorescent quencher dye at the 3′ end. The probe is cleaved as the polymerase extends from the PCR primer, separating the reporter dye and the quencher dye and resulting in an increase in fluorescence intensity.

***Taq* polymerase** a polymerase that is thermostable to 95°C used in polymerase chain reaction amplification reactions; isolated from *Thermus aquaticus*.

taxonomic/taxonomy the scientific classification of microorganisms involving identification and nomenclature.

TCDD 2,3,7,8-tetrachlorodibenzodioxin.

$TCID_{50}$ tissue culture infectious dose 50.

TC_{LO} lowest concentration reported to result in an adverse health effect in humans or test animals.

TCR Total Coliform Rule.

t_d generation time.

TD_{LO} lowest dose reported to cause toxic effects in humans or test animals.

T-DNA a segment of the Ti plasmid of *Agrobacterium* that is transferred to plant cells.

TDT thermal death time.

teichoic acid complex polymers of polyols and phosphate that are associated with bacterial cell walls and with the cytoplasmic membrane of gram-positive bacteria; serve as antigens and used to define antigenic groups of some genera.

teleomorph term used in mycology to describe the sexual state.

Teliomycetidae fungal subclass; yeasts, rusts, and smuts within the Basidiomycetes.

teliospore thick walled, binucleate resting spore; dry, powdery airborne stage of smuts; difficult to distinguish microscopically from myxomycetes.

TEM transmission electron microscopy.

temperate phage/temperate virus a virus that does not destroy its host bacterial cell; viral nucleic acid is incorporated into the host cell's genome and replicated with the host bacterial cell without causing lysis, a process called lysogeny.

temperature gradient gel electrophoresis (TGGE) an analytical method in which gel electrophoresis, which separates molecules on the basis of size, electrical charge, and conformation, is combined with a temperature gradient (perpendicular to the electrical gradient) that separates the molecules on the basis of their thermal stability.

template a physical structure used as a model for replication of an identical unit; a nucleotide sequence that directs the synthesis of a complementary sequence of nucleotides.

ten-degree temperature quotient Q_{10} value.

tenuous delicate.

teratogen/teratogenic capable of causing birth defects in offspring.

terminal at the end.

terminal electron acceptor the compound at the end of a series of oxidation/reduction reactions that becomes reduced as a result of the gaining of electrons.

terminal restriction fragment (TRF) segment of DNA used to construct a profile for determining microbial communities; terminal restriction fragment length polymorphism.

terminal restriction fragment length polymorphism (T-RFLP) molecular method using high resolution gel electrophoresis with an automated DNA sequencer to rapidly compare the diversity of microorganisms in environmental samples by comparing the variation in the length of fluorescently labeled terminal restriction fragments of DNA sequences.

terpenoid toxin a toxin that has a $(C_5H_8)_n$ configuration; includes trichothecenes and fusidanes.

terricolous living on soil.

tertiary treatment processing of water or wastewater beyond secondary treatment; for potable water the additional processing is designed to minimize the presence of pathogenic microorganisms, generally consists of sequential processing using sedimentation, coagulation, and disinfection; for wastewater the additional processing is designed to remove non-biodegradable organic pollutants and mineral nutrients, especially nitrogen and phosphorus, generally consists of activated carbon filtration; may also involve more advanced treatment methods (e.g., ozonation, reverse osmosis) if the wastewater is to be recycled for beneficial purposes.

tetracycline an antibiotic that inhibits protein synthesis and is more effective in intact prokaryotic cells than eukaryotic cells.

Tetrahymena pyriformis protozoan species; ciliated phagocytic parasite found in natural and drinking waters worldwide that serves as a natural host and reservoir for members of the MAIS group of *Mycobacterium* spp.

T-even phage bacteriophage in the Myoviridae family of viruses; contain double-stranded DNA, characterized by a complex structure consisting of a modified icosahedral head and contractile tail.

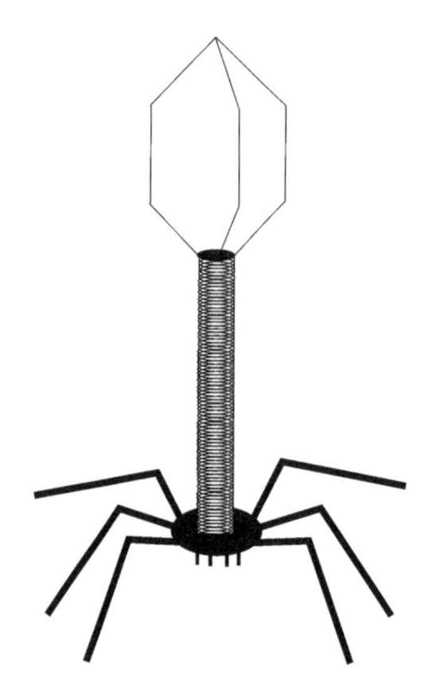

TFF tangential flow filtration.

TGGE temperature gradient gel electrophoresis.

thallic the transformation of a segment of fungal hyphae into conidia by the formation of crosswalls.

thallus the vegetative and reproductive structures that make up a fungal colony.

theoretical model mathematical relationships used to describe fundamental or general ecological principles; in contrast to a simulation model.

thermal death time (TDT) the amount of time required at a particular temperature to kill a specified number of organisms.

thermocline thermal region of water that is characterized by a rapid decrease in temperature and little mixing.

Thermococcus bacterial genus; member of the *Archaea* that are characterized as motile, obligate anaerobic, chemoorganotrophic, hyperthermophilic cocci that are indigenous to anoxic submarine thermal waters and grow on proteins and some sugars using S° as the electron acceptor in the reduction to H_2S.

Thermococcus kodakaraensis bacterial species; growth range of 65–95°C with an optimal growth temperature of 90°C.

thermocycler instrument to control the temperature and reaction time involved in the polymerase chain reaction amplification.

Thermodesulfobacterium bacterial genus; characterized as group I non-acetate oxidizing thermophilic dissimilatory sulfate reducing bacteria with an optimal growth temperature of 70°C.

Thermodesulforhabdus bacterial genus; characterized as group II acetate oxidizing dissimilatory sulfate reducing bacteria that are thermophilic, motile, gram-negative bacilli that utilize fatty acids up to C18.

Thermodiscus bacteria genus; member of the Archaea; obligate chemoorganotrophic organisms that complex organic mixtures using S° as the electron donor and are thermophilic with a growth range of 75–98°C and an optimal growth temperature of 90°C, but are not able to grow above 100°C.

thermophile, thermophilic descriptive of the temperature requirements of microorganisms with an optimal growth temperature greater than 40°C.

thermophilic actinomycetes gram-positive, filamentous spore-forming bacteria that grow at elevated temperatures (>40°C) and mimic fungi with branching structures; inhalation exposure to some species is associated with respiratory distress in humans.

Thermoplasma bacterial genus; member of the order Mollicutes; characterized as aerobic, chemoorganotrophic, thermophilic, and acidophilic with an optimal growth temperature of 55°C and pH of 2; lipoglycan constitutes the major fraction of their total lipid.

Thermoplasma acidophilum bacterial species; mostly isolated from coal refuse piles that are self-heating due to spontaneous combustion; one isolate from acid hot springs.

thermotolerant having the ability to withstand elevated temperatures.

Thermus aquaticus bacterial species; thermophilic organism that serves as the source of *Taq* polymerase.

thin-layer chromatography (TLC) analytical chemistry method for the separation and identification of compounds using a silica-coated plate; samples are spotted onto the plate and a solvent is applied resulting in the migration of compounds up the plate.

thin sections/thin sectioning preparation of specimens for viewing with electron microscope in which glass fragments are used as knives to slice sequential segments of a sample.

Thiobacillus bacterial genus; gram-negative rod-shaped cells that oxidize sulfur and sulfur compounds to sulfate; widely distributed in soils, fresh water, and seawater.

Thiobacillus ferrooxidans bacterial species; in addition to the oxidation of sulfur and sulfur compounds, this organism can oxidize ferrous iron to ferric iron.

Thiobacillus thiooxidans bacterial species; metabolic products result in acid mine drainage; isolated from waste coal heaps.

Thiobacillus thioparus bacterial species; an autotrophic bacterium that oxidizes thiocyanate to ammonia and CO_2 using the released sulfide as an electron donor and energy source; formerly named *Thiobacillus thiocyanooxidans*.

thioglycolate broth liquid medium used to assess the oxygen requirements of bacteria.

Thiospira bacterial genus; spiral-shaped cells with pointed ends that oxidize sulfur and sulfur compounds; found in water overlying sulfur-rich mud.

thiosulfate a chemical routinely used in the water industry to neutralize the effects of chlorine.

Thiothrix bacterial genus; a chemolithotroph that oxidizes hydrogen sulfide and grows only as a mixotroph.

Thiovulum bacterial genus; round to ovoid cells that oxidize sulfur and sulfur compounds and often have sulfur inclusions in their cytoplasm; found in fresh and sea water.

THMs trihalomethanes.

threshold limit value (TLV) concentration established by the American Conference of Governmental Industrial Hygienists as a concentration of substances and conditions under which it is believed that nearly all workers may be repeatedly exposed day after day without adverse health effects.

thymine a pyrimidine base that is one of the four nucleotides comprising DNA molecules; thymine forms a bond with adenine on the opposite strand of the DNA molecule.

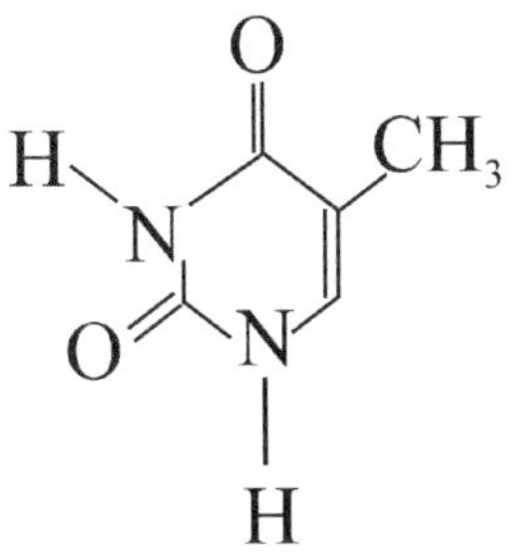

thyrsiform having middle branches that are longer that those above or below.

tickborne transmission transfer of a disease-causing organism by the bite of a tick as the vector.

time-weighted average (TWA) a term used in industrial hygiene to describe average allowable concentrations of exposure for an 8 or 10 hour period of time.

Ti plasmid an *Agrobacterium* plasmid that is responsible for the transfer of genes from bacteria to plants.

tissue culture the maintenance and growth of pieces of tissue in culture using appropriate nutrients; commonly referred to as cell culture, in which cells derived from tissues are maintained and grown in culture; used as host cells for virus or protozoan propagation and detection.

tissue culture infectious dose 50 ($TCID_{50}$) the number of viruses or parasites required to produce infection in 50% of the tissue culture flasks inoculated.

TLC thin-layer chromatography.

TLV threshold limit value.

TM7 newly described proposed bacterial division in which no organisms have yet been isolated in pure culture but have been characterized by molecular methods.

TMDL total maximum daily load.

TMV tobacco mosaic virus.

tobacco mosaic virus (TMV) causative agent of the leaf disease of tobacco and tomato plants; one of earliest studied viruses, it was demonstrated as a filterable agent in 1892 by Ivanowsky and having properties of a living organism in 1899 by Beijerinck.

toluidine blue a dye used with light microscopy for the visualization of fungal hyphae and spores which appear deep red in plant tissue samples that stain a blue or greenish blue.

tomentum an entangled layer of long, soft fibrils.

top-fermenting yeast a brewery yeast, generally *Saccharomyces cerevisiae*, that remains uniformly distributed in the fermenting liquid and is transferred to the surface by CO_2 generated during the processing, in contrast to bottom yeast.

Torula fungal genus; dark brown powdery colonies with aerial hyphae and warty brown blastoconidia formed in branching chains; ubiquitous; 8 species; isolated from soil, grasses, sugar beet roots, ground nuts and oats, dead herbaceous plant stems, jute, wicker, straw baskets, and cellulose-containing building materials; associated with Type I allergies.

Torula herbarum fungal species with cytotoxic properties.

toruloid/torulose the successive swelling and constriction of some fungal hyphae or conidiophores.

total coliform/total coliform bacteria aerobic and facultative anaerobic, gram-negative, non-spore-forming rod-shaped bacteria that ferment lactose with the production of gas within 48 hours at 35°C; includes *Citrobacter* spp., *Klebsiella* spp., *Enterobacter* spp., and *Escherichia coli*.

total coliform membrane filter (MF) method standardized procedure specified in the Safe Drinking Water Act for the analysis of total coliform bacteria in drinking water by water utilities; the method involves the filtration of a 100 ml water sample onto a filter and incubation on eosin methylene blue (EMB) agar; results may not exceed 1 CFU/100 ml as the arithmetic mean of all samples examined per month, 4 CFU/100 ml in more than 1 sample when <20 samples are examined/month, or 4 CFU/100 ml in >5% of the samples when >20 samples are examined/month; results are reported to the United States Environmental Protection Agency and if the data do not meet the prescribed standard the water utility must notify the public and correct the problem.

Total Coliform Rule (TCR) a regulation promulgated by the United States Environmental Protection Agency in 1989 (which became effective in 1990) that requires all community water systems and noncommunity water systems to monitor their water for the presence of total coliform bacteria.

total count measurement of both culturable and nonculturable organisms in a sample, generally obtained using microscopic assay or electronic particle counting.

total maximum daily load (TMDL) a calculation of the maximum amount of a pollutant that a water body can receive from all sources (point and nonpoint), and still meet water quality standards for the intended use(s) of that water body; the program is mandated by the Clean Water Act and administered by the United States Environmental Protection Agency.

total risk the sum of the likelihood of developing disease resulting from exposure risk and background risk.

toxigenicity the severity of a toxin.

toxin a chemical produced naturally as a by-product of metabolism by a microorganism that will damage other organisms.

toxoid a nontoxic chemical similar in structure to a toxin that is used to induce an antibody response.

T phages one of a group of seven bacteriophages that infect *Escherichia coli*; the T-even phages (T2, T4, and T6) cause host cell metabolism to cease upon infection, while the odd-number T phages depend on the continuation of host cell metabolism after infection for their replication.

T-2 toxin trichothecene mycotoxin; listed by International Agency for Research on Cancer with limited evidence for animal carcinogenicity but inadequate evidence for human carcinogenicity; detected in conidia of *Fusarium graminearum* and *Fusarium sporotrichioidies*.

trace elements chemical constituents that are present in low concentrations.

tracer the use of a chemical, radioactive element, or a microorganism to determine the migration of a solution or if a sample has been contaminated during collection; often used in mapping of ground water and in the collection of subsurface soil samples.

tracheitis inflammation of the trachea.

transcription the synthesis of RNA using a DNA template.

transduction genetic recombination process mediated by virus; DNA is incorporated from one cell to another with the assistance of a virus by generalized transduction or specialized transduction.

transfer ribonucleic acid (tRNA) the form of RNA that is used to carry amino acids to the ribosome; the ribosomes then link the amino acids together to form proteins.

transformation genetic recombination in bacteria following uptake of DNA that was free in the environment.

transgenic microorganism a microorganism with a cloned DNA sequence from another organism.

transient occurring for only a short, discrete period of time.

transient noncommunity water system a noncommunity water system that serves a changing population of people once or a few times per year; examples include campgrounds, highway rest stops and parks.

translation the synthesis of protein using mRNA as a template and tRNA as an adaptor.

translocation a mutation involving a change in location of a large section of chromosomal DNA.

translucent allowing light to pass through, but diffusing the light so that objects are not completely distinguishable; in contrast to transparent.

transmission spread of a disease-causing microorganism.

transmission electron microscopy (TEM) an electron microscopic technique used to view the internal structure of thin sections of microorganisms using electrons and electromagnets under vacuum.

transparent clear; allowing light to pass through; in contrast to translucent.

transparent tape sampling method used to assess fungal contamination on smooth surfaces with analysis by light microscopy; in contrast to tape lift used for assay of fungal colonies in laboratory culture.

transposable element transposon.

transposable phage the bacteriophage Mu that is capable of replicating as a transposon.

transposon small DNA sequences that are mobile and can replicate with insertion into sites within a chromosome.

trench method a means of refuse disposal in a sanitary landfill in which waste is deposited into a trench and covered over with soil extracted from an adjacent area forming a trench for the next day's deposit; in contrast to the ramp variation and the area method.

TRF terminal restriction fragment.

T-RFLP/TRFLP terminal restriction fragment length polymorphism.

tricarboxylic acid cycle citric acid cycle.

Trichoderma fungal genus; rapidly spreading white floccose colony on malt extract agar developing tufts of blue-green to yellow-green spores in a slimy mass when incubated in diffuse daylight; spores are similar in size and shape to those of *Aspergillus* and *Penicillium* when viewed with light microscopy, but a distinctive green pigment can be used for discrimination; colorless reverse; ubiquitous; approximately 20 recognized species; isolated from soil, decaying wood and woodchips, grains, citrus fruit, vegetables, wallpaper, paper, tapestry, textiles, and unglazed ceramics; production of cellulases that have been associated with allergic reactions among bakers; associated with Type I allergies and Type III hypersensitivity, and considered an emerging opportunistic pathogen in the immunocompromised; potential for production of trichothecenes.

Trichoderma harzianum fungal species; conidia are smooth-walled; produces antifungal compounds that are effective in protecting plants from pathogens so it is incorporated in ground bark used to protect trees and vegetable crops.

Trichoderma viride fungal species; conidia are rough-walled; produces a ketone metabolite with a distinctive coconut odor.

trichothecenes terpenoid mycotoxins; considered to be the most potent small molecule inhibitors of protein synthesis; mode of action is the inhibition of peptidyltransferase activity; cross the blood/brain barrier and have a variety of effects on neural transmission; can cause alveolar macrophage defects and may effect phagocytosis; several toxins within this grouping such as T-2 toxin and *Fusarium* toxin; possible uses in biological warfare.

trickling filter an aerobic secondary wastewater treatment process in which decomposition is accomplished using a bed of crushed rock or other solid support matrix approx. 2 m thick over which the sewage is applied, by a revolving sprayer or sprinkler; organic material is oxidized by microorganisms in biofilms that develop on the rock or solid support matrix.

trifurcate three-forked in appearance.

trihalomethanes (THMs) any one of a group of chemicals that result from the chlorination of waters containing organic materials; many of these compounds are known or suspected carcinogens.

triplate segmented dish designed for the use of three different culture media.

tripoint inoculation deposition of a pure culture suspension onto three discrete points of an agar surface.

tRNA transfer ribonucleic acid.

trophic level an individual step in a food chain or food web involved in the transfer of energy stored in organic compounds.

trophozoite vegetative stage of a protozoan.

true positive result that correctly identifies the presence of the analyte of interest and does not falsely record its presence when it is absent.

truncate cut-off at right angles; descriptive of the abrupt end at the base of some fungal cells.

trypticase soy agar tryptic soy agar.

trypticase soy broth tryptic soy broth.

tryptic soy agar (TSA) a general purpose medium containing organic compounds and agar used for the culture and isolation of a variety of bacteria and fungi.

tryptic soy broth (TSB) a general purpose liquid medium containing organic compounds used for the culture of a variety of bacteria and fungi.

TSA tryptic soy agar.

TSB tryptic soy broth.

t-test Student's t-test.

tube dilution serial dilution.

tubercle a wart-like protuberance.

tuberculate having wart-like or knob-like projections.

tuberculosis human disease, usually involving the lungs, caused by exposure to *Mycobacterium tuberculosis*; disease is spread by the aerosolization of infectious bacteria expelled from the lungs or throat of infected persons who cough or sneeze.

tuberous rounded and swollen.

tularemia disease caused by exposure to *Francisella tularensis*; characterized by the sudden onset of fever, chills, headache, muscle aches, and weakness; may progress to pneumonia; transmitted to humans by arthropod bites, handling carcasses of infected animals, inhaling infected aerosols, or ingestion of contaminated food or water.

tumor abnormal mass of tissue that may or may not be malignant.

tumorgenic capable of causing tumors.

turbidity measurement assessment of the density of microorganisms or particles in a liquid suspension using a photometer or spectrophotometer; the amount of light scattering is used to calculate the density of the cells or particles with the greater the density resulting in increased turbidity.

turbinate shaped like a top.

turgid rigid or filled out due to high fluid content.

TWA time-weighted average.

Tween a surfactant used as an amendment to buffers and aqueous solutions to assist in the suspension of particulate material.

Twort, F.W. discoverer of bacterial viruses in 1915 independently from d'Herelle.

two-tailed test comparison of any difference between groups regardless of the direction of the difference; in contrast to a one-tailed test.

Tyndall, John (1820–1893) English physicist who demonstrated in 1877 that particle-free air does not cause spoilage of sterilized substances by using discontinuous heat treatments (tyndallization); credited with Cohn for the initial experimentation with endospores as heat-resistant structures.

tyndallization sterilization technique in which heat is applied in short sequences of steam.

Type I allergies an immediate adverse reaction to an antigen (e.g., allergic asthma and hay fever).

Type III allergies an immune complex reaction resulting from the deposit of antibody-antigen complexes in affected tissue.

Type IV delayed hypersensitivity a cell mediated, delayed antibody-dependent reaction involving T-lymphocytes and macrophages that results in the formation of a granuloma at the site of exposure.

Type I error the error of concluding that a difference existed when there was no difference; the probability of this conclusion is denoted as α or alpha error.

Type II error the error of concluding that no difference existed when there was a difference; the probability of this conclusion is denoted as β or beta error.

type strain archived specimen of a microorganism used for taxonomic comparison.

U

ubiquitous widespread occurrence.

Ulocladium fungal genus; colony on malt extract agar is granular to velvety in texture, dark to rusty brown in color with geniculate sporulating structures that are visible when viewed with a stereomicroscope; spore fragments, short chains, and young spores may be confused with *Alternaria* or *Pithomyces*; ubiquitous; approximately 9 recognized species; isolated from soil, dung, grasses, fibers, wood, decaying plant material, paper, textiles, paint, gypsum board, tapestries, jute and other straw materials; associated with Type I allergies and evidence that the major allergen is Alt a I; cross reacts with *Alternaria*.

ultracentrifuge instrument used to separate molecules in suspension using centrifugation at high g forces.

ultrafiltration a hydraulic pressure-driven method used to remove, separate, or recover particles and colloids from a stream of liquid; the pore size and molecular configuration of the filter are factors in determining the size of the retained particles.

ultrafreezer storage device for materials at temperatures below –20°C.

ultrasonic disruption the use of vibrations ≥20 kHz to break apart microbial cells in a liquid suspension as a result of pressure changes that are produced with alternating development and collapse of gas bubbles.

ultraviolet (UV) radiation generally considered as wavelengths between 220–300 nm; does not penetrate solid, opaque, light-absorbing materials but can be used for disinfection of exposed surfaces; effective as a biocidal agent by affecting the purine and pyrimidine bases of DNA and RNA; exposure may result in mutagenesis of survivors with the formation of pyrimidine dimers.

umber dark brown or reddish brown in color.

umbo a raised knob or mounded in the center.

undulant wavy in appearance.

undulate having a wavy-like motion.

unicellular single cell.

uniseriate in one line or one series; descriptive of the genus *Aspergillus* with the formation of the phialides directly on the vesicle without metulae.

United States Environmental Protection Agency (US EPA) a regulatory agency of the federal government of the United States with the mission to protect human health and to safeguard the natural environment; organized into 10 regional offices; some state, county, and local governments also use the designation EPA for their respective agencies.

universal tree of life phylogenetic tree.

upper detection limit (UDL)/upper limit of detection largest quantity that can be accurately determined.

upper limit of quantitation/upper quantitation limit highest number that can be accurately enumerated.

upper respiratory tract the nose, mouth, sinuses, and pharynx.

upstream primer the primer that binds to the 5′ end of the target sequence.

uracil a pyrimidine base that is one of the four nucleotides comprising RNA molecules.

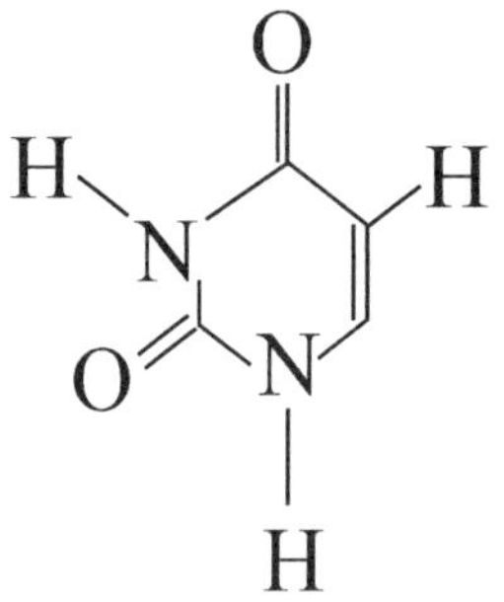

Urocystis fungal genus; member of the Basidiomycotina; causative agent of smut of onion.

Ustilago fungal genus; member of the Basidiomycotina; parasitic on seeds and flowers of cereals and grasses; causative agent of smut corn and wheat; isolated from respiratory tract of humans resulting from the inhalation route of exposure; in culture the colonies are white, slow growing, moist, and compact becoming velvety or powdery with age; cells are irregular yeast-like or elongated and spindle-shaped.

UV ultraviolet radiation.

V

vacuole a large, fluid-filled sac located in the cytoplasm.

vacuum pump a mechanical device that supplies a vacuum (i.e., an environment in which the air pressure is below 1 atmosphere); used for operation of equipment that draws air or liquid through an orifice.

vacuum sampling the collection of settled particulate using a vacuum pump; see settled dust sampling.

V8 agar fungal culture medium with a water activity of 0.98; used to increase sporulation.

validity the ability of a test to measure an observation; the ability of an assay to distinguish between the presence and the absence of an analyte of interest.

valine ($C_5H_{11}NO_2$) an essential amino acid that cannot be synthesized by the human body and therefore must be ingested in foods.

valley fever coccidiomycosis.

Vampirococcus name given to predatory bacteria that use the cytoplasmic constituents of other bacteria as a nutrient source.

van der Waals forces electromagnetic forces at the atomic level.

van Ermengem, E.M.P. discoverer of *Clostridium botulinum* as the causative agent of botulism in 1896.

van Leeuwenhoek, A. Leeuwenhoek, Antonie van.

van Niel, C.B. a student of Kluyver; used the enrichment culture technique to isolate and study phototrophic and chemolithotrophic bacteria with a focus on the photosynthetic processes of sulfur bacteria; brought the Delft School tradition to the United States for the training of general bacteriologists.

vapor the gaseous phase of a chemical that is a solid or a liquid at room temperature.

vapor pressure the measure of the tendency of a liquid to form a gas with the higher the value the more volatile the liquid; the measurement is a function of temperature; values reported in millimeters of mercury (mmHg).

variable number of tandem repeats (VNTR) sequences of repetitive DNA; used in DNA fingerprinting.

variables what is being measured or observed; generally classified as a dependent variable or independent variable.

variance (σ^2, S^2) measure of the average deviation of individual values.

$$s^2 = \frac{\Sigma (X-\overline{X})^2}{n-1}$$

Varicella-zoster virus a member of the *Varicellavirus* genus of the Alphaherpesvirinae subfamily of the Herpesviridae family; relatively large (150–200 nm diameter), enveloped, icosahedral double-stranded DNA viruses; cause of chickenpox and shingles in humans.

VB broth Vogel-Bonner broth.

VBNC viable (but) non-culturable.

vector an animate object that aids in the indirect transmission of an infectious agent to a new host; generally subdivided into biological vector and mechanical vector categories; alternative usage in biotechnology, see cloning vector.

VEE Venezuelan equine encephalitis.

vehicle a nonliving source of indirect transmission of a pathogen to a new host; commonly associated with food and water that results in disease in a large number of people; also used to describe the transport of pharmaceuticals to a site in an infected host using a microorganism.

velutinous velvety in appearance.

Venezuelan equine encephalitis (VEE) virus a member of the *Alphavirus* genus (also referred to as group A arboviruses) of the Togaviridae family; 60–70 nm diameter, icosahedral, single-stranded RNA viruses; transmitted by mosquitoes; causes a rarely-fatal form of encephalitis in humans; see also Western equine encephalitis virus, Eastern equine encephalitis virus.

ventricose swollen or enlarged in the middle.

Venturia inaequalis fungal species; phytopathogen, causative agent of apple scab.

verdigris-green bluish green in color.

vermiform worm-shaped.

verruciform wart-like in appearance.

verrucose having small rounded wart-like ends.

verticil/verticilate the formation of conidiogenous cells on a conidiophore in a pattern resembling the spokes of a wheel.

Verticillium fungal genus; hyaline or brightly colored conidia borne in slimy heads, some in chains; numerous saprophytic species and some phytopathogens resulting in wilt.

Verticillium dahliae fungal species; phytopathogen of strawberry, oilseed rape, and potato.

vesicle a swollen sac-shaped structure descriptive of some fungal conidiophores and sporangiophores.

VFF vortex flow filtration.

viable/viability physiological state in which an organism can maintain life and reproduce.

viable nonculturable (VNC) a physiological state in which an organism can respire, but is incapable of growth on solid culture media.

viable plate count plate count assay.

Vibrionaceae bacterial family; characterized as gram-negative, facultatively anaerobic bacilli that are either non-motile or motile by means of polar flagella.

Vibrio bacterial genus; member of the Vibrionaceae; characterized as gram-negative, facultative anaerobic, oxidase positive bacilli with a fermentative metabolism; generally found in aquatic environments.

Vibrio anguillarum bacterial species; causative agent of fish disease, commonly occurring in hatcheries.

Vibrio cholerae bacterial species; water-borne transmission with the consumption of contaminated oysters and seawater; causative agent of cholera in humans.

Vibrio fischeri bacterial species; luminescent bacterium that produces *N*-β-ketocaproylhomoserine lactone, an autoinducer to produce bioluminescence.

Vibrio parahaemolyticus bacterial species; marine organism commonly isolated from seawater, shellfish, and crustaceans; ingestion route of exposure resulting in gastroenteritis, especially prevalent in Japan where raw fish is consumed.

vibriosis a primary septicemia of captured, maintained eels caused by *Vibrio vulnificus*.

Vibrio vulnificus bacterial species; estuarine gram-negative pathogen that is a primary pathogen for eels and may cause severe infections in humans.

villose a hairy or shaggy soft hair appearance.

vinaceous wine-colored.

vinegar product resulting from the conversion of ethyl alcohol to acetic acid by acetic acid bacteria generally by members of the genera *Acetobacter* and *Gluconobacter*; wine or apple cider are often used as starting materials although pure alcohol mixed with water results in a product labeled distilled vinegar.

***virA* gene** segment of the Ti plasmid of *Agrobacterium tumefaciens*; genetic sequence that codes for a protein kinase that interacts with signal molecules and then phosphorylates the product of the *virG* gene.

***virB* gene** segment of the Ti plasmid of *Agrobacterium tumefaciens*; genetic sequence that codes for a protein that facilitates the transfer of T-DNA in the form of a bridge membrane and a pore-like structure.

***virD* gene** segment of the Ti plasmid of *Agrobacterium tumefaciens*; genetic sequence that codes for an endonuclease that nicks DNA in the Ti plasmid adjacent to the T-DNA.

***virE* gene** segment of the Ti plasmid of *Agrobacterium tumefaciens*; genetic sequence that codes for a single-stranded DNA-binding protein that binds single strands of T-DNA and transfers them into the plant cell.

***virG* gene** segment of the Ti plasmid of *Agrobacterium tumefaciens*; gene sequence that activates transcription of other *vir* genes.

***vir* genes** genetic sequences on the Ti plasmid that code for proteins that are essential for T-DNA transfer; gene expression is induced by molecules synthesized by wounded plant tissue.

viricidal/virucidal having the ability to cause a virus to become incapable of infection.

virion a single structurally mature, intact, virus particle.

viroid small RNA molecule with activity that mimics virus infectivity.

virosorb filter, 1MDS a 0.45 μm porosity, positively charged membrane filter that adsorbs negatively charged viruses and is used to concentrate enteric viruses in water.

virulent capable of causing disease.

virulent phage lytic phage.

virus a small (generally 20–300 nm), acellular biological entity which possesses a nucleic acid genome and replicates as an obligate intracellular parasite within a living host cell; classical viruses have a nucleic acid genome surrounded by a protein shell nucleocapsid that may be surrounded by additional layers which may include proteins and lipids; metabolically inactive outside its host.

virus-like particles (VLPs) particles resembling viruses that are produced using molecular biology procedures; one or more features of the virus particle, such as the nucleic acid or portions of the virus necessary for pathogenicity, is not present; also a term used to describe particles observed in fungi using transmission electron microscopy that appear like viruses, but have no known effect or lytic mechanism.

virustatic having the ability to interfere with the replication of a virus.

visible light radiation relatively narrow band in the electromagnetic spectrum ranging from 320–800 nm.

vital stains dyes that are incorporated by living cells; used to distinguish viable from nonviable cells.

VLPs virus-like particles.

VNA viral nucleic acid.

VNC viable nonculturable.

VNTR variable number of tandem repeats.

VOC volatile organic compound.

volatile readily vaporized at room temperature.

volatile organic compound (VOC) organic compound that is readily vaporized at room temperature.

volumetric pipet calibrated pipet for the transfer of a specified amount of a liquid; in contrast to a serological pipet.

volunteer bias referral bias.

vortex table-top device used for mixing of solutions in a test tube using a swirling motion of variable intensity.

vortex flow filtration (VFF) a high efficiency ultrafiltration method used to collect virus in water samples on to a filter matrix that incorporates bypassing large volumes of water prior to concentration to avoid the presence of other suspended solids in the sample.

VOSC volatile organic sulfur compound.

Waksman, Selman (1888–1973) Russian-born American scientist; discoverer of streptomycin.

Wallemia xerophilic fungal genus; colony powdery in texture and chocolate brown in color; cylindrical conidiophores with elongated conidiogenous cells that subdivide into 4 cube-shaped conidia; isolated from soil, air, dry food products, grains, house dust, textiles, and paper; isolated on agar medium with low water activity such as dichloran-18% glycerol-agar (DG18).

Wallemia sebi fungal species; slow growing fan-like or stellate, powdery colonies that are orange-brown to blackish brown in color with cylindrical, pale brown, smooth-walled conidiophores; osmophilic airborne fungus isolated from forest soil, textiles, timber, paper, pecan nuts, salted fish and beans, and stored hay and may cause ulcerative abscesses in humans; cultured on high osmotic medium.

warm vent a hydrothermal vent that emits temperatures of 6–23°C.

wastewater effluent the liquid portion of the product of wastewater treatment.

wastewater treatment a series of biological and physical processes designed to remove particulate to reduce the biological oxygen demand, decrease the concentration of nutrients (e.g., nitrogen and phosphorus), and remove or inactivate pathogens in domestic sewage prior to disposal.

water activity (Aw, a_w) measurement of water availability; the vapor pressure of water in substrate divided by the vapor pressure of pure water; values vary from 0 to 1 with pure water having an Aw of 1.

$$Aw = \frac{\text{vapor pressure of water in substrate}}{\text{vapor pressure of pure water}}$$

water balance diffusion of water into and out of a cell.

waterbath a water-holding container used in the laboratory that provides a uniform temperature.

waterborne transmission transfer of microbial contaminants via water; generally an ingestion route of exposure that primarily results in intestinal disease although dermal contact of parasites in water can result in disease.

water content the amount of water naturally present in a material.

water holding capacity (WHC) the maximum amount of moisture that can be retained in a substance, usually used in studies with soil.

water molds aquatic fungi that are members of the Oomycetes.

water purification a series of processes that are designed to remove a specified percent of various microorganisms from water.

water quality descriptive of the microbiological and chemical content of water to minimize the risk of waterborne disease transmission.

waterwashed disease illness caused by organisms that originate in feces and are transmitted through contact because of inadequate sanitation.

Watson, James presented the model of the structure of DNA with Francis Crick in 1953.

weapons of mass destruction biological warfare agents and radiological agents used to evoke terror and illness in large populations of people.

weathering physical, chemical and biological processes that assist in the formation of soils.

WEE Western equine encephalitis.

WEELs workplace environmental exposure limits.

weft a felt-like mat of fungal hyphae.

Western blot a biotechnology detection method involving protein-antibody binding, also termed immunoblot; in contrast to Southern blot or Northern blot hybridization procedures.

Western equine encephalitis (WEE) virus a member of the *Alphavirus* genus (also referred to as group A arboviruses) of the Togaviridae family; 60–70 nm diameter, icosahedral, single-stranded RNA viruses; transmitted by mosquitoes; infection may cause a mild fever or more serious illness with central nervous system involvement; see also Eastern equine encephalitis virus, Venezuelan equine encephalitis virus.

wet mount preparation of a sample for microscopic analysis in which a specimen is suspended in a liquid droplet.

wet rot a condition in which the composition of wood is changed as a result of saturation with water, often occurring with wood that is submerged in water or water-logged soil.

WHC water holding capacity.

wheat rust disease caused by basidiomycetes.

whey a waste liquid of the dairy industry containing lactose and minerals that is used in industrial processes as supplemental carbon.

white rot fungi category of basidiomycetes that degrade both the lignin and cellulose in wood; in contrast to the brown rot fungi.

Whittaker, Robert H. proposed the five-kingdom classification system for living organisms in 1969.

whorl a group of branches that radiates from a common point.

wild type an organism isolated from nature.

wilt droopiness of plant structures due to the loss of turgor; symptom of some plant diseases caused by microorganisms.

windrow long, generally rectangular, pile of material in a static pile composting facility.

Winogradsky, Sergei (1856–1953) Russian scientist who described microbial oxidation of hydrogen sulfide and sulfur in 1887, and ferrous sulfide in 1888, discovered chemolithothophy while studying colorless sulfur bacteria, *Clostridium pasteurianum*, ananaerobic nitrogen-fixing bacterium in 1893, originated the nutritional classification of soil bacteria, and is considered as the founder of soil microbiology.

Woese, Carl proposed a three-kingdom classification system consisting of Archaebacteria, Eubacteria, and Eukaryotes in the 1980s.

wood-rotting fungi referring to basidiomycetes; commonly classified as brown rot and white rot fungi.

working distance the range of distance where the image formed by a lens is clearly focused.

workplace environmental exposure limits (WEELs) airborne exposure limits for chemicals established by the American Industrial Hygiene Association.

w/v weight/volume.

x symbol used to denote an unknown quantity.

X used to denote the value for one data point.

$\bar{X}$ symbol for arithmetic mean.

xanthomonadins yellow, membrane-bound, halogenated aryl polyene pigments that are produced by *Xanthomonas* spp. and may provide some protection against photodamage.

Xanthomonas bacterial genus; member of the family Pseudomonaceae; chemoorganotrophic, gram-negative, straight, obligate aerobic, bacillus that is motile by a single polar flagellum; many species are phytopathogens.

Xanthomonas campestris bacterial species; phytopathogen, causative agent of black rot of crucifers.

Xanthomonas citri bacterial species; phytopathogen, causative agent of canker of citrus.

Xanthomonas hyacinthi bacterial species; phytopathogen, yellow pigmented organism, causative agent of yellow disease in bulb flowers and is readily spread by wind, rain, and bulb injury.

Xanthomonas maltophila bacterial species; not a phytopathogen; potential environmental exposure to humans via aerosols generated during wastewater treatment practices.

Xanthomonas oryzae bacterial species; phytopathogen, causative agent of blight of rice.

Xanthomonas phaseoli bacterial species; phytopathogen, causative agent of blight of beans.

Xanthomonas pruni bacterial species; phytopathogen, causative agent of leaf spot of fruits.

Xanthomonas vascularum bacterial species; phytopathogen, causative agent of gumming of sugar cane.

x-axis the horizontal axis of a graph, the abscissa; in contrast to the y-axis.

xenobiotic a synthetic product that does not occur naturally in the environment.

xerophile/xerophilic descriptive of microorganisms that are capable of living in a dry environment; organisms with an optimal water activity ranging from 0.65 to 0.90.

XPS X-ray photoemission spectroscopy.

X-ray diffraction (XRD) an analytical technique used for the identification of crystalline material including products produced during corrosion.

X-ray photoemission spectroscopy (XPS) an analytical technique used in characterizing corrosion products that identifies atomic species based on electron core level binding energies.

XRD X-ray diffraction.

y-axis the vertical axis of a graph, the ordinate; in contrast to the x-axis.

yeast unicellular fungus that reproduces by budding; most are members of the Ascomycetes; some have a filamentous phase; for example, *Saccharomyces cerevisiae* are used extensively in food production for leavening of bread and in beer and wine fermentation.

yeast extract broth used as collection medium for airborne *Legionella*.

Yersin, A.J.E. discoverer of Yersinia pestis as the causative agent of the plague in 1894.

Yersinia bacterial genus; member of the family Enterobacteriaceae; facultative, gram-negative, nonsporulating bacilli with 10 established species, most with simple nutritional requirements; previously classified as the genus *Pasteurella*.

Yersinia enterocolitica bacterial species; causative agent of an intestinal infection in humans and warm-blooded wild and domestic animals; also isolated from soil, surface water, reptiles, fish and shellfish; ingestion route of exposure with contaminated food and water.

Yersinia pestis bacterial species; causative agent of bubonic plague; 200 mammalian species naturally infected but rodents are the principal natural reservoir with the bacterium residing in the alimentary canal including the salivary glands and stomach, transmission by flea bite; organisms enter the bloodstream of the host and are transported to the nearest lymph node where they amplify resulting in swelling; organisms may also cause a secondary septicemia, meningitis or pneumonia; transmitted via aerosols from pneumonic plague patients; classified as a biosafety level 3 agent.

Yersinia pseudotuberculosis bacterial species; causative agent of a tuberculosis-like disease in the lymph nodes of animals but only rarely in humans; may cause diarrhea, septicemia and mesenteric lymphadenitis in humans; main reservoirs are rodents, hares and rabbits, and wild birds; organism survives for long periods in soil and river water.

Yo Pro-1 a nucleic acid-staining fluorescent dye used with epifluorescent microscopy that has been shown to provide a brighter fluorescence than 4′, 6′-diamidino-2-phenylindole (DAPI) staining.

Z

z, Z standard score.

zearalenone a patented growth stimulant for animals and an anabolic steroid; produced by *Fusarium culmorum*.

zee score standard score.

zed score standard score.

Zenorhabdus bacterial genus; symbionts of entomopathogenic nematode family Steinernematidae which participate in maintaining suitable conditions for nematode reproduction; pathogenic to insects resulting from toxigenic reaction and septicemia.

zero-order epidemic microscale spread of plant disease limited to a few hundred meters within one field and within one growing season; in contrast to first-order epidemic and second-order epidemic.

zeta potential a measure of the electrostatic potential, determined by the presence of charged groups on the surface of a cell or molecule.

Ziehl-Neelsen stain a staining technique using carbofuchsin with gentle heat, acid-alcohol de-colorization and a methylene blue counterstain for the detection of Mycobacterium spp. with light microscopy.

Zinder, Norton discoverer of bacterial transduction.

ZoBell, Claude considered the founder of marine microbiology.

ZoBell sampler apparatus used for the manual collection of water samples in lakes and reservoirs below the surface.

zonate arranged in zones or rings radiating from the center.

zone of inhibition the area in which an antimicrobial substance prevents growth.

zooglea referring to a mass of bacteria and/or lower algae that are held together by a slime matrix and water.

Zooglea ramigera bacterial species; produces extracellular polysaccharide slime matrix during sewage treatment.

zoonosis an animal disease transmitted from animal to animal and from animals to humans.

zoophilic prefers animals rather than humans or soil.

z-score standard score.

zwitterion having both acid and basic portions; characteristic of amino acids.

Zygomycetes fungal class; rapid growing non-septate fungi with sporangiospores produced in sporangia and some species have rhizoids and stolons; sexual reproduction produces a dark, thick-walled zygospore.

zygomycosis diseases caused by members of the Zygomycetes.

Zygomycotina a subdivision in the Amastigomycota; saprophytic, parasitic, or predatory fungi that are characterized by the formation of a zygospore; Mucorales and Trichomycetes are two of the subdivisions of Zygomycotina.

zygospore the sexual spore produced by members of the Zygomycotina that results from the fusion of two gametangia.

Zygosporium fungal genus; conidia are colorless without a distinctive morphology; isolated from damp interior walls, dead leaves and soil; no information available on human health effects, toxicity, or allergenicity.

zymocide a factor present in some yeast cells that is toxic to other yeasts.

zymogenous opportunistic microorganisms found in the soil.

zymology the study of yeast.

Zymomonas bacterial genus; tolerant of low pH and ethanol concentrations up to 10%; large, gram-negative bacillus that ferments sugars to ethanol; active in fermentation of plant sap for industrial production (e.g., fermentation of agave in Mexico to produce tequila and palm sap in tropical areas); responsible for spoilage of fruit juices and production of an odor of rotten apples in spoiled beer.